SpringerWienNewYork

Konrad Soyez
Hartmut Graßl

Climate Change and Technological Options

Basic facts, Evaluation
and Practical Solutions

SpringerWienNewYork

Priv.-Doz. Dr.-Ing. habil. Konrad Soyez
Potsdam University
Chair on Vegetation Ecology and Nature Conservation
Potsdam, Germany

Prof. Dr. Hartmut Graßl
Max Planck Institute for Meteorology
Hamburg, Germany

Das Werk ist urheberrechtlich geschützt.
Die dadurch begründeten Rechte, insbesondere die der Übersetzung, des Nachdruckes, der Entnahme von Abbildungen, der Funksendung, der Wiedergabe auf photomechanischem oder ähnlichem Wege und der Speicherung in Datenverarbeitungsanlagen, bleiben, auch bei nur auszugsweiser Verwertung, vorbehalten.

© 2008 Springer-Verlag/Wien
Printed in Germany

SpringerWienNewYork is part of Springer Science Business Media
springer.at

Product Liability: The publisher can give no guarantee for all the information contained in this book. This does also refer to information about drug dosage and application thereof. In every individual case the respective user must check its accuracy by consulting other pharmaceutical literature. The use of registered names, trademarks, etc. in this publication does not imply, even in the absence of a specific statement, that such names are exempt from the relevant protective laws and regulations and therefore free for general use.

Typesetting: Karson Grafik- und Verlagsservice, 1020 Vienna, Austria
Printing: Strauss GmbH, 69509 Mörlenbach, Germany

Printed on acid-free and chlorine-free bleached paper

With 59 (partly coloured) figures and 48 tables
SPIN: 12205172

Library of Congress Control Number: 2008931583

ISBN 978-3-211-78202-6 SpringerWienNewYork

Table of contents

Table of contents V
List of Abbreviations XI
1 Introduction 1
2 The Climate System 3
    2.1 Climate System Components 3
    2.2 Observed Climate Variability and Change.................................... 4
    2.3 The Greenhouse Effect of the Atmosphere 7
    2.4 Greenhouse Gases 7
    2.5 The Carbon Cycle and Atmospheric Carbon Dioxide................. 10
    2.6 Aerosols, their Direct and Indirect Climate Effects 12
        2.6.1 Direct aerosol particle effects 12
        2.6.2 Indirect aerosol particle effects... 14
    2.7 Radiative Forcing of Climate Change 15
    2.8 Physical Climate Processes and Feedbacks 17
        2.8.1 Solar and terrestrial radiation 18
        2.8.2 Clouds.. 18
        2.8.3 Intensity of the meridional overturning in the Atlantic 20
        2.8.4 Shift of extratropical storm tracks 21
        2.8.5 Soot versus cloud condensation nuclei 24
    2.9 Atmospheric Chemistry and Climate 24
        2.9.1 Stratospheric ozone depletion... 24
        2.9.2 Photochemical smog or increased tropospheric ozone 25
    2.10 Climate Change and Vegetation... 26
        2.10.1 Climate impact of reforestation in the boreal zone........ 27
        2.10.2 Climate impact of reforestation or deforestation in the tropics 27
3 Climate Modelling.. 29
    3.1 Model Basics and Structure 29
    3.2 Climate Model Evaluation 31
    3.3 Emission Scenarios 32
    3.4 Projections of Climate Change.. 33
    3.5 Regional Climate Change Information 38
        3.5.1 Why is regional climate modelling needed?................... 38

4	Consequences of Mean Global Warming		43
	4.1	Shrinking of the Cryosphere	43
	4.2	Changes in Sea Level	44
	4.3	Changed Precipitation Distribution	45
	4.4	Detection of Climate Change and Attribution of Causes	46
		4.4.1 What is detection of anthropogenic climate change?	47
		4.4.2 What is an attribution of anthropogenic climate change?	47
		4.4.3 First detection of anthropogenic climate change	48
		4.4.4 Attribution of climate change to causes	49
5	Impacts of and Adaptation to Climate Change		51
	5.1	Vulnerability	51
	5.2	What is a Climate Change Impact?	52
		5.2.1 Impacts on sectors	53
		5.2.2 Impact on certain geographical regions	56
6	Sustainable Development and Climate Change		59
	6.1	The Enhanced Greenhouse Effect without Analogues in Climate History	59
	6.2	Carbon Cycle Feedbacks	60
	6.3	Sequestration of Carbon	61
	6.4	Barriers, Opportunities and Market Potential for New Technologies and Practices	62
		6.4.1 Basis of opportunities: The technical potential of renewable energy	63
		6.4.2 Development status of renewable energy sources	64
		6.4.3 Emissions Trading as a New Practice	65
	6.5	Costs of Adaptation	66
	6.6	Mitigation of Climate Change as the Prerequisite for Sustainable Development	66
		6.6.1 Guardrails	67
		6.6.2 Geo-engineering?	68
	6.7	A Sustainable Energy Path	69
7	International Climate Policy Approaches		71
	7.1	First Policies, Measures and Instruments	71
		7.1.1 The Villach Conferences	71
		7.1.2 Intergovernmental Panel on Climate Change (IPCC)	72
	7.2	United Nations Framework Convention on Climate Change	72
	7.3	From Rio to Kyoto	73
		7.3.1 Kyoto Protocol	74
	7.4	Climate Protection Goals in Europe and Germany	75

		7.4.1	Emission Reduction Goals and Measures in Germany	76
		7.4.2	Emission Reduction Goals and Measures in the European Union	76
		7.4.3	Reduction Goals	78
	7.5	Sustainable Development Strategy (of the EU)		78
8	Evaluation of climate effects			81
	8.1	The general evaluation problem		81
	8.2	Life Cycle Assessment for climate control		85
		8.2.1	Background issues	85
		8.2.2	LCA Methodology	87
			8.2.2.1 Step 1: Definition of goal and scope	87
			8.2.2.2 Step 2: Life-cycle inventory analysis	88
			8.2.2.3 Step 3: Life Cycle Impact Assessment	89
			8.2.2.4 Step 4: Interpretation	91
		8.2.3	LCA case study: Comparison of climate effects of integrated waste management systems	92
			8.2.3.1 Definition of goal and scope	92
			8.2.3.2 Technology description and functional unit	93
			8.2.3.3 Impact assessment results	94
			8.2.3.4 Interpretation and conclusions	95
9	Climate effects and mitigation potentials of economic sectors			97
	9.1	Processes and typical emissions		99
	9.2	General mitigation potentials		100
	9.3	Mitigation potential by Carbon Capture and Storage (CCS)		103
10	Climate impacts and emission mitigation of industrial production			107
	10.1	Relevance und trends of industrial sector emissions		107
	10.2	Conseqences of climate change for industry		111
	10.3	Climate impacts and emission mitigation of selected industrial processes		112
		10.3.1	Production of Iron and Steel	112
		10.3.2	Cement and lime manufacture	114
			10.3.2.1 Cement manufacture	114
			10.3.2.2 Lime manufacture	116
		10.3.3	Ammonia manufacture and urea application	117
		10.3.4	Aluminum production	118
		10.3.5	Carbon Dioxide Use	119
		10.3.6	Semiconductor manufacture	120
		10.3.7	Nitric and adipic acid production	121
			10.3.7.1 Nitric acid	121
			10.3.7.2 Adipic acid	121
11	Climate effects of agricultural processes			123

	11.1	Overview on agriculture and climate interaction............................ 123
	11.2	Greenhouse gas emissions by livestock.. 126
	11.3	Climate effect of manure management and biogas.......................... 127
		11.3.1 Manure management.. 127
		11.3.2 Biogas production 129
		11.3.3 Case study: Agricultural biogas production in a sports and recreation center... 131
	11.4	Rice cultivation... 133
	11.5	Agricultural soils 134
12		Climate effects of waste management 137
	12.1	Background 137
	12.2	Source reduction and waste recycling 139
		12.2.1 Background and preconditions.. 139
		12.2.2 GHG effects of source reduction and recycling............... 140
		12.2.2.1 Source reduction effects............................... 140
		12.2.2.2 Waste material recycling effects 142
		12.2.3 Case study: GHG effects of the German packaging material recycling system DSD...................................... 144
	12.3	Composting 147
		12.3.1 Composting process characters.. 147
		12.3.2 GHG sources in composting 148
		12.3.3 Carbon sequestration by compost application.................. 151
		12.3.4 Use of composting CO_2 as greenhouse fertilizer 152
	12.4	Climate effects of waste deposition in landfills................................ 152
		12.4.1 Climate effects by landfill gas emissions.......................... 153
		12.4.1.1 Overview on landfill gas generation 154
		12.4.1.2 Landfill gas recovery 155
		12.4.1.3 Effects of landfill management 157
		12.4.2 Carbon storage by solid waste deposition......................... 157
	12.5	Climate effects of waste combustion.. 158
		12.5.1 Technological background – Waste to Energy (WtE) 158
		12.5.2 Climate effects by MSW combustion 159
		12.5.2.1 CO_2, N_2O and pollutant emissions............ 159
		12.5.2.2 Beneficial climate effects by recovery....... 160
	12.6	Climate effects of mechanical-biological waste pre-treatment... 161
		12.6.1 Technological background... 161
		12.6.2 Climate effects by MBP technology 164
		12.6.2.1 Climate effects of material recovery 165
		12.6.2.2 Optimising the climate impacts of the MBP technology 167
		12.6.2.3 Climate effects of MBP waste gas treatment 168

Table of contents

		12.7	WARM – a tool for GHG evaluation of waste management strategies 169
13	Energy related climate impacts 175
	13.1	GHG emission overview... 176
	13.2	GHG emissions by extraction of fossil fuels............................ 178
		13.2.1	Methane emissions by oil and natural gas extraction... 178
		13.2.2	Methane emissions from coal mines............................ 179
	13.3	Climate effects of fossil fuel use 181
		13.3.1	Climate effects of fuel use in power production.......... 181
		13.3.2	Climate effects of transportation 184
			13.3.2.1 Road transportation 184
			13.3.2.2 Aircraft transportation 186
	13.4	Climate effects of other non-biomass energy sources............... 187
	13.5	Climate effects of biomass derived fuels 189
		13.5.1	Biofuels – facts and definitions 190
		13.5.2	Current policies promoting biofuels 193
		13.5.3	Budgeting of climate consequences of biofuels 196
			13.5.3.1 Climate impacts of bioethanol................... 196
			13.5.3.2 Climate impacts of biodiesel 200
		13.5.4	Consequences for biofuels application 201
14	Individual activities to reduce climate impacts 203
	14.1	Climate impacts of production and consumption of goods 203
	14.2	Climate impacts of modern city life style 204
	14.3	Climate oriented individual behaviour...................................... 205

Literature 209

Indication of Sources in Subtitles of Figures... 219

List of Abbreviations

Al	Aluminum
AP	Acidification potential
AR4	Forth Assessment Report of IPPC
AT_4	Respiration activity coefficient
BtL	Biomass-to-liquid
CCS	Carbon Capture and Storage
CDM	Clean Development Mechanisms
CDP	Carbon Disclosure Project
CFC	Chlorofluorohydrocarbons
CH_4	Methane
CO	Carbon monoxide
CO_2	Carbon dioxide
CO_2-eq.	Carbon dioxide equivalents
COP	Conference of Parties
Day_{Emit}	Methane emission rate
d.m.	Dry matter
DME	Dimethylether
DSD	Duales System Deutschland
EC	European Commission
EIA	Environmental Impact Assessment
EJ	Exajoule; 10^{18} J
EMIC	Earth model of intermediate complexity
EOR	Enhanced oil recovery
EPA	U.S. Environmental Protection Agency
ETBE	Ethyl-Tertio-Butyl-Ether
EU	European Union
EU-15	European Union (15 member states)
EU-23	European Union (23 member states)
FCCC	Framework Convention on Climate Change
Fe	Iron
FGD	Flue gas desulfurisation
GB_{21}	Gas generation coefficient

GDP	Gross Domestic Product
GE	Gross energy intake
GHG	Greenhouse gas
GNP	Gross National Product
Gt	Gigatons; 10^{12} kg; 10^9 t
GWP	Global Warming Potential
HDPE	High density polyethylene
HEF	Hydrofluoroether
HFC	Fluorinated hydrocarbons
ICSU	International Council of Scientific Unions
IPCC	Intergovernmental Panel on Climate Change
ISO	International Organization for Standardization
JAMA	Japan Automobile Manufacturer Association
JI	Joint Implementation
K	Kelvin
KAMA	Korean Automobile Manufacturer Association
kg	Kilogramm
LCA	Life cycle assessment
LDPE	Low density polyethylene
LFG	Landfill gas
LLDPE	Linear low density polyethylene
MBP	Mechanical-biological pre-treatment of residual waste
Mio	Million; 10^6
MPI	Max Planck Institute
MSW	Municipal solid waste
MTBE	Methyl-Tertio-Butyl-Ether
NADW	North Atlantic Deep Water
NEI	National Emission Inventory
NMVOC	Non methane volatile organic carbon
NP	Nutrition potential
OECD	Organisation for Economic Co-operation and Development
ODP	Ozone depletion potential
PCC	Precipitated calcium carbonate
PCOP	Photochemical oxidation potential
PECVD	Plasma enhanced chemical vapour deposition
PET	Polyethylenetherephthalat
PFC	Perfluorocarbon

List of Abbreviations

ppb	Parts per billion
ppm	Parts per million
RDF	Refuse derived fuel
RFI	Radiative Forcing Index
RFS	Renewable Fuels Standard
RTO	Regenerative thermal oxidation
SAR	Second Assessment Report
SETAC	Society of Environmental Toxicology and Chemistry
SF_6	Sulfur Hexafluoride
SRES	Special Report on Emission Scenarios
SWCC	Second World Climate Conference
t	Ton; 10^3 kg
TAR	Third Assessment Report of IPPC
TOC	Total organic carbon
UNCED	United Nations Conference on Environment and Development
UNEP	United Nations Environmental Programme
UNFCCC	United Nations Framework Convention on Climate Change
VEETC	Volumetric Ethanol Excise Tax Credit
WARM	Waste Reduction Model
WBGU	German Global Change Advisory Council
WCRP	World Climate Research Programme
WMO	World Meteorological Organisation
WSSD	World Summit on Sustainable Development
WtE	Waste to energy

1 Introduction

Climate change has been considered a fact for over a decade, following the proof of rising CO_2 levels, rising Earth's temperatures, melting of glaciers, etc. The consequences can be observed in many regions in daily life, by events such as more frequent or stronger flooding of rivers, increased storms and snowfall, cloudbursts, as well as drought, and desertification. The reason for climate change, natural or anthropogenic, has been under discussion for a long time.

There is no doubt that mankind contributes to climate change through activities connected with emissions of climatically relevant gases. For example, use of fossil fuels with high emissions of carbon dioxide and other climate gases, especially in transportation and traffic, industrial production and application of substances (which are climate gases of extreme high warming potentials), agricultural activities (such as animal husbandry and rice cultivation) leading to emissions of methane or nitrous oxide, and methane emissions from landfills caused by ineffective waste management, etc.

To control the situation, reduction measures of climate gases and other relevant actions, are urgently necessary on all levels. This is understood by the public and by policy makers. Climate related activities thus are high ranking on the political agenda. They are implemented into the political programmes on UN level, internationally and single countries, but also on communal levels by climate initiatives of cities or NGOs. Examples are the so-called Kyoto Protocol reducing climate gas emissions in industrialized countries, bans of halogenated hydrocarbons, shifting of energy sources from fossil to renewable and international CO_2-emission trading. It is but obvious, that the efforts must be strengthened, to reduce the risks of a dangerous interference with the climate system.

This publication is intended to give a more detailed insight into the problem and the efforts to tackle it, so that the reader is able to develop best climate strategies in practical cases in his own field. In the first part, fundamentals of climate and climate change are discussed, including facts on climate system functioning and its modelling, as well as the effect of climate change on sustainable development and international policy approaches. In the second part, climate and related effects of technology are evaluated, and an overview of effects of industrial and agricultural processes on climate are given, including technological, infrastructural, economic, and socially oriented activities to reduce climate gas emissions.

2 The Climate System

Climate is one of the most important natural resources. Given the size of the planet Earth and its mean distance from the Sun, the three leading climate parameters are solar energy flux density, clouds plus precipitation and land surface characteristics. Asking for the most fundamental parameters for our life we get a very similar answer: energy from the sun, water from the skies and photosynthesis of plants. Hence, climate determines where we can live in larger numbers, what food we get, and how we have to protect ourselves against weather related extremes.

It is therefore obvious that decision makers have to deal with climate whether it is changing because of external forcing or just varying because of internal interaction of climate system components. It has become common practice to speak of the climate system and its components in order to point to its complexity because of the manifold interactions within components and among them. As box 2.1 underlines the components interact at very different timescales from minutes to billions of years, thus creating continuous changes of climate, to which all living beings have to adapt but to which also all living beings have contributed.

2.1 Climate System Components

All parts of the Earth system are important for the climate in a certain area. Therefore, no single place is independent of all others on our globe. The best example of the component interactions are the joint glacials and interglacials of both hemispheres, although the triggering comes from the northern hemisphere with its major landmasses. If the northern hemisphere is closest to the Sun in boreal winter, the declination of the Sun is high (it can vary from 21.8 to 24.5°) and the eccentricity of the Earth's orbit is larger than on average the probability for an inception of a glacial is high. As observations of atmospheric composition, reconstructed from air bubbles in Antarctic ice, confirm, the main long-lived greenhouse gases in the atmosphere, namely carbon dioxide (CO_2), nitrous oxide (N_2O) and methane (CH_4), then start declining and also the southern hemisphere with higher solar radiation flux density will "descend" into a glacial as well. Although the processes leading to this joint glacials are not yet fully understood, it is clear that ocean, atmosphere, biosphere and cryosphere have interacted strongly to create global mean surface temperatures 5°C lower than in interglacials.

> **Box 2.1:** **Climate System Components and their Typical Timescales**
>
> Climate must change continuously as the components of the climate system interact non-linearly at very different time scales from minutes to billion years (see also table 2.1). And because the radiation flux density of the Sun also varies on time scales from minutes to billions of years. In addition, the Earth's orbit around the Sun varies quasi-periodically through changed positions of the neighbouring major planets (Venus, Jupiter, Saturn). Understanding the Earth system, of which climate is an important part, is therefore a very complex endeavour and far from being in a very mature stage.

Table 2.1: Climate system components and their typical time scales, together with some climate phenomena

Component	Typical Time-scales	Some Climatically Relevant Phenomena
Atmosphere	Minutes to Millennia	Planetary Boundary Layer Height, Greenhouse Gas Composition, Annual Cycles of Temperature and Precipitation, Storm Tracks
Ocean	Seasons to about 100,000 years	Boundary Currents (e.g. Gulf Stream), Global Conveyor Belt, Single and Multi-year Sea Ice
Biosphere	Days to Millennia	Blooming, Biomass Production, Biome Distribution, Vegetation Cover, High Biodiversity in the Tropics, Vegetation Period, Anthropogenic Monocultures, Algae Blooms, Food Webs in Ocean and on Land
Cryosphere	Days to Millions of Years	Snow Cover, Ice Sheets, Ice Caps, Mountain Glaciers, Permafrost, Frozen Ground, Lake and River Ice, Sea Ice
Lithosphere	Years to Many Million Years	Continental Drift, Subduction of Oceanic Crusts, Formation of Mountain Ranges, Earthquakes, Volcanoes, Fossil Fuel Formation
Pedosphere (Soils)	Decades to Many Millennia	Weathering of Rocks, Humus Formation, Cycling of Elements through Microbiological Processes, Changed Atmospheric Composition by Emissions from Soils

2.2 Observed Climate Variability and Change

One of the most obvious characteristics of climate is its variability, especially in areas with strong gradients of climate zones, e.g. in the semi-arid tropics and in higher mid-latitudes. The mean temperature of one of the coldest days in July and one of the warmest in December in Hamburg do not differ. The rain in parts of the Northern Sahel from one year to the next may differ by more than a factor 3. Therefore climate – as the synthesis of weather – is not only characterized by averages of parameter values but also by their frequency distributions (see figure 2.1 and box 2.2). Although strong deviations from the average value are rare, they get most of the attention because they represent weather extremes to which our infrastructures are often not well adapted. As Figure 2.1 also clarifies, new extremes on the side to which the distribution is shifted must accompany climate change. The only exception would be the case with a strongly narrowing frequency distribution (what has not been observed). Do we already observe manifestations of climate change? Yes, there are numerous ones, besides the obvious mean global near surface air temperature increase over the recent 150 years (see figure 2.2). These are (only examples):

- Accelerated mean retreat of mountain glaciers worldwide,
- Strong decrease of multi-year sea ice in the Arctic Ocean (-7 percent per decade since 1979 when satellite observations began),
- Reduced snow cover over North America, less pronounced over Eurasia,
- More rain per event in nearly all areas with slightly decreasing, constant or increasing total precipitation,
- Reduced daily temperature amplitude which can be caused by higher water vapour content, increased cloudiness and higher atmospheric turbidity,
- Mean global sea level rise, about 1.5 to 2.0 mm/a in the 20th century, recently increased to ~3 mm/a, as observed by satellite altimeter measurements since 1991,
- Increased yearly precipitation in most high latitude areas, decreased yearly precipitation in the semi-arid subtropics,
- Decreased temperatures in the stratosphere and mesosphere,
- Increased vegetation period length (about 2 weeks) in the Northern Hemisphere higher latitudes,
- Changed optical properties of clouds caused by air pollution.

In the present rapid climate change era single evaluations of very long time series are therefore partly misleading as the recent decades might have shown a changed frequency distribution of a climate parameter, e.g. rain amounts per event. Therefore subsections of long time series have to be evaluated separately.

Strong climate variability on time-scales up to millennia, as oceans and ice sheets are involved in creating it, makes it difficult to separate climate variability from real climate change. Observations are always the result of variability and change. For more details see section 4.4.

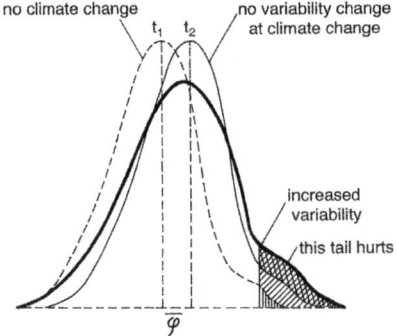

Figure 2.1: Schematic frequency distribution of climate parameters both for present climate and changed climate. Also a broadened distribution for a changed climate is shown (Grassl, 2002).

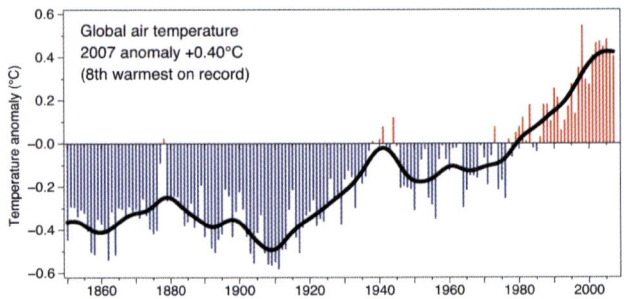

Figure 2.2: Global mean near surface air temperature since 1856 (Meteorological Office of the United Kingdom)

Box 2.2: Climate, its Variability and Change

The World Meteorological Organization (WMO) in Geneva, a Specialized Agency of the United Nations, defines climate as the synthesis of weather extracted from frequent atmospheric and surface parameter observations over at least 30 years. This synthesis must contain the probability of deviations from the mean, i.e., the number of events deviating for example 3 standard deviations σ from the mean must be known. If mathematical functions are fitted to the observations the frequency distributions turn into probability density functions. If the frequency distribution is close to a Gaussian distribution, like for temperature, a 3 σ-event is close to what most would call a hundred year event, observed only once per century. However, for such long periods the frequency distribution might have changed when climate change has occurred.

Climate change occurs if external parameters change, like solar radiation flux density. It has become customary to speak of climate change also if volcanic eruptions reach the stratosphere or mankind changes atmospheric composition inadvertently, although we and volcanoes are part of the Earth system.

2.3 The Greenhouse Effect of the Atmosphere

If the transmission of solar radiation to the surface of a planet is less attenuated than the emission of thermal (heat) radiation from the surface to space, the surface and the lower atmosphere of the planet warm until emission at the top of the atmosphere balances (in a multi-year average) the amount of solar radiation absorbed by the atmosphere and the surface. For the planet Earth the warming at the surface because of strong absorption of thermal radiation by some atmospheric gases is caused nearly exclusively by minor constituents, i.e. by less than three per mille of the mass of the atmosphere. The warming caused by these minor constituents is about 30°C. The use of the word "about" is due to the fact that we do not know which reflectivity for solar radiation the Earth's surface would have without these gases, as the ocean might not exist as at present. Since the heat absorbing gases act like glass covering a greenhouse, the warming effect caused by them is called in the rough analogy greenhouse effect.

2.4 Greenhouse Gases

The main constituents of the atmosphere are nitrogen (N_2) with 78.09 percent, oxygen (O_2) with 20.94 percent and Argon (A) with 0.93 percent, constituting already 99.96 percent of the dry atmosphere. The minor constituents absorbing strongly thermal infrared (heat) radiation are (if ranked according to importance in the undisturbed pre-industrial atmosphere):

1. Water vapour (H_2O), responsible for nearly two-thirds of the greenhouse effect;
2. Carbon dioxide (CO_2), responsible for about 20 percent;
3. Ozone (O_3), responsible for about 7 percent;
4. Nitrous oxide (N_2O), contributing only about 3 percent;
5. Methane (CH_4) contributing less than 3 percent.

Some other naturally occurring gases, like carbon monoxide (CO), are weak greenhouse gases but are neglected here.

It is important to note that two of these five greenhouse gases are short-lived, namely water vapour and ozone, with lifetimes of about 9 days and hours to months, respectively. The other three gases are all called long-lived, although they differ strongly in their lifetimes. Methane needs about 12 years and nitrous oxide 120 years until their concentration would have fallen to about 37 percent, i.e. 1/e, if no emissions occurred. For carbon dioxide no single number can be given as the uptake into the ocean is a complex process that involves several time scales, e.g. sedimentation of organisms. About 200 years are needed to reach 1/e for the additional (anthropogenic) load.

Table 2.2   Recent changes of naturally occurring long-lived greenhouse gases due to human activities (IPCC, 2007a)

Species	Concentration		
Year	1750	2005	Change since 1998
CO_2 (ppm)	280	379 ± 0.65	+ 13
CH_4 (ppb)	730	$1,774 \pm 1.8$	+ 11
N_2O (ppb)	270	319 ± 0.12	+ 5

The assessment of the consequences of an enhanced greenhouse effect (for concentration changes see table 2.2.) is made more complicated by the strong temperature dependence of the water cycle including the dominant greenhouse gas water vapour (see box. 2.3). Radiative transfer calculations with fixed atmospheric composition, except a doubling of carbon dioxide concentration, and allowing so-called convective adjustment in the troposphere, give a 1.2°C average warming of near surface air temperature. If water vapour reacts, like in so-called equilibrium models of general atmospheric circulation, the warming roughly doubles mainly due to two positive feedbacks, one by water vapour (already mentioned) and the other by the snow/ice-albedo[1]/temperature feedback. The big uncertainty still remaining for the sensitivity of the climate system to an enhanced greenhouse effect is due to the less well known feedback of clouds, where even the sign of the global mean effect is not known, although locally it is clear that more low, optically thick water clouds would

---

[1] Albedo is the ratio between backscattered and incoming solar radiation flux density. Therefore it is zero for a black body and unity for a completely backscattering or reflecting surface. Typical values of natural surfaces like forests and grassland vary from about 10 to 20 percent, but can reach 90 percent for fresh powder snow.

Greenhouse Gases

dampen the enhanced greenhouse effect (negative feedback) and that cold but thin cirrus (ice clouds) in the upper troposphere would enhance it.

Box 2.3   Known Positive Feedbacks in the Water Cycle

The dominant cycle for the climate system is the water cycle. This dominance is due to several positive and potentially also negative feedbacks. Two positive feedbacks have to be named here caused by:

- The Clausius-Clapeyron equation
- The differences between the albedo of snow/ice and liquid water

Feedback 1: Assuming chemical equilibrium and the second law of thermodynamics we get for the change of water vapour pressure in the atmosphere dp_s at saturation for a temperature change dT

$$\frac{dp_s}{dT} \approx \frac{L}{vT^2}$$

with

L = latent heat of evaporation
v = specific volume of water vapour
T = absolute temperature (K)

For atmospheric temperatures between +25°C at saturation (base of a tropical cumulus cloud) and -80°C at saturation (tropical cirrus cloud top) the saturation pressure p_s increases from 6 to 20 percent per °C temperature change following this equation. Hence the water vapour pressure varies by up to four orders of magnitude between cloud base and cloud top of a severe tropical cumulonimbus. In other words: Nearly all water vapour in this column (~ 60 mm of precipitable water) will fall out as rain, for a convergent flow even more.

Therefore, the dominant greenhouse gas water vapour will show a positive feedback (i.e. it will amplify) when a warming is stimulated by a greenhouse gas concentration increase.

Feedback 2: The brightest and the darkest natural surface are composed of water: Fresh powder snow "reflects" (in reality mainly backscatters) about 85 percent of incoming solar radiation flux density while the ocean absorbs about 96 percent at blue skies and high sun; hence reflecting only 4 percent. Consequently the disappearance of snow or sea ice warms the lower atmosphere which leads to further melting nearby. This positive feedback is the key feedback in glacial cycles and is still important at present, as large parts of the Northern Hemisphere land and ocean are seasonally and smaller parts permanently covered by snow or snow on sea ice. This positive so-called snow/ice-albedo/temperature feedback is also fundamental for the inception of a glacial.

2.5 The Carbon Cycle and Atmospheric Carbon Dioxide

Besides the water cycle also the carbon cycle is fundamental for the Earth's climate. All Earth system components contain carbon and exchange it rapidly or slowly causing major changes of climate. The most active part of the carbon cycle is the exchange between the atmosphere and the biosphere, both on land and in the sea. As figure 2.3 shows the reservoir atmosphere with about 760 GtC looses about 120 ± 70 GtC per year because carbon dioxide (CO_2) is

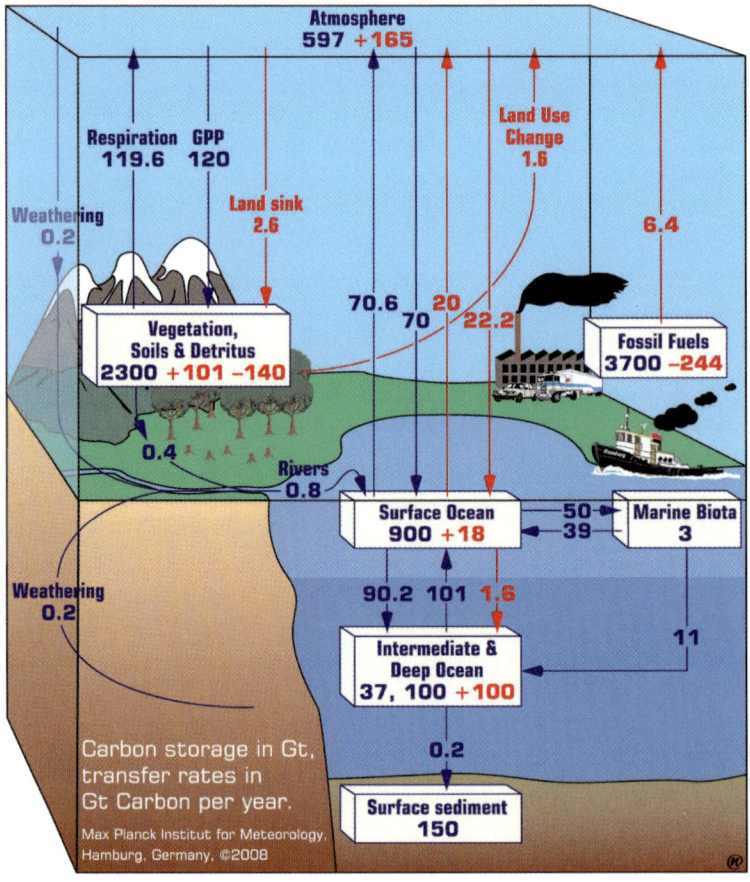

Figure 2.3: Carbon fluxes between the reservoirs in GtC and GtC/a, respectively (IPCC, 2007a)

The Carbon Cycle and Atmospheric Carbon Dioxide

taken up by terrestrial plants or enters the sea, where part of it is used by algae for biomass production. Nearly the same amount is going back to the atmosphere because of out-gassing from the ocean and respiration by plants, animals and humans. Thus, the undisturbed carbon cycle is nearly balanced, as only 0.2 GtC/a are buried in sea sediments and a still not well known small portion (0.4 GtC/a) enlarges peat bogs or is emanating from or buried in deep soils. From the above we learn that the lifetime of carbon dioxide in the atmosphere is rather short, about seven years, and the question arises "Why is there a problem with additional CO_2 as "only" about 8 GtC/a, roughly one percent of the reservoir content, are emitted by human activities?" As Table 2.3 clarifies 4.1 ± 0.1 GtC/a of these emissions remain in the atmosphere, hence increase the atmospheric carbon reservoir by about half a percent per year. Only 2.2 Gt of the anthropogenic carbon enter the ocean per year and thus are taken away for hundreds of years. Would we stop emissions only about 15 percent of all former anthropogenic emissions will in the long-run stay in the atmosphere. Table 2.3 contains a further astonishing fact: In the 1990s the existing terrestrial biosphere grew, by about 2.6 ± 1.7 GtC/a, more than the emissions from deforestation and other land use change that reached 1.6 ± 1.1 GtC/a. Overall, the carbon stored in the biosphere increased in recent years.

How long will this favourable situation last? No answer can be given yet but coupled carbon cycle and climate models, forerunners of the emerging Earth system models, show at least that there is the possibility for a sign change in the latter part of the 21st century. Then climate change impacts are much stronger and the carbon dioxide fertilization effect draws down carbon dioxide levels in the atmosphere less as climate change puts more pressure on terrestrial and marine ecosystems, which then reduce the CO_2-uptake. The lesson from this section: the net fluxes into the ocean interior count and not gross primary production on land or fluxes from the atmosphere across the ocean surface. While the carbon cycle was nearly balanced before industrialization, it is strongly imbalanced now with a "hangover" of more than 3 GtC/a for the atmosphere.

Table 2.3: Present anthropogenic carbon emissions and heat fluxes; all in GtC/a (IPCC, 2007a)

	Period		
	1980s	1990s	2000 – 2005
Atmospheric increase	3.3 ± 0.1	3.2 ± 0.1	4.1 ± 0.1
Emissions (fossil + cement)	+5.4 ± 0.3	+6.4 ± 0.4	+7.2 ± 0.3
Net flux ocean/atmosphere	-1.8 ± 0.8	-2.2 ± 0.4	-2.2 ± 0.5
Net flux land/atmosphere	-0.3 ± 0.9	-1.0 ± 0.6	-0.9 ± 0.6
Land use change flux	+1.4 (0.4 to 2.3)	+1.6 (0.5 to 2.7)	n.a.
Residual terrestrial sink	-1.7 (-3.4 to 0.2)	-2.6 (-4.3 to -0.9)	n.a.

2.6 Aerosols, their Direct and Indirect Climate Effects

Although greenhouse gases and cloud droplets and cloud ice are dominant constituents of the atmosphere determining to a large extent the radiation budget of the planet Earth, also aerosol[2] particles play a major role for our planet. These tiny particles suspended in air in the size range from about 1 nanometer (nm) to about 10 micrometers (μm) radius are either emitted from the Earth's surface or form in the atmosphere from precursor gases like sulphur dioxide (SO_2). While the small ones (< 0.01 μm radius) often get attached to other larger particles or surfaces by the molecules' Brownian motion, the larger ones, called coarse particles, with r > 1μm, settle through gravity. Therefore, maximum spectral concentration (particles per unit volume per unit of radius) is often close to r = 0.01 μm and typical lifetime in the free troposphere reaches weeks for particles in the size range around 0.1 μm. Their main sink process is activation as a cloud condensation nucleus and subsequent rain-out, and much less below cloud scavenging. As any cloud droplet needs an aerosol particle as a condensation nucleus, it is clear that the aerosol particles can strongly influence the optical properties of clouds and can exert an indirect effect besides the direct one, which anybody can see through atmospheric turbidity, the result of scattering of visible light by aerosol particles.

2.6.1 Direct aerosol particle effects

All aerosol particles scatter, absorb and emit radiation. Depending on their size extinction (scattering plus absorption) of solar radiation is often much more important than emission of thermal infrared radiation. Often the latter is neglected, which is only valid for particles with radii < 0.1 μm. Especially when the particles have grown with relative humidity, this neglection is not justified. As most of them are soluble and hygroscopic, i.e. they deliquesce like sodium chloride (NaCl) at about 80 percent relative humidity, turbidity of air increases strongly at high relative humidity. Hence attenuation or extinction of solar radiation by aerosol particles is not only depending on chemical composition but also on relative humidity. Via air pollution we thus change radiative transfer in the cloud free atmosphere. Main effects are:

---

[2] An aerosol is a mixture of air and small particles suspended in air. Often the term aerosol is used for the particles only. Here the term aerosol particles will be preferred.

Reduced solar radiation flux density at the surface:

It may reach 50 Wm^{-2} in many metropolitan areas and thus more than compensate increased thermal infrared back radiation of the atmosphere due to the enhanced greenhouse effect of the atmosphere.

Mostly enhanced local planetary albedo

The word mostly is needed as the effect of scattering and absorbing aerosol particles on the albedo at the top of the atmosphere is also depending on the albedo of the surface. An example: Sulfate particles with soot attached to some over a snow covered area may not change at all local planetary albedo but would strongly enhance it, if moved over an open water surface. On global scale estimates of the overall aerosol particle effect fall into the -0.5 Wm^{-2} range, thus enhancing albedo through more backscattering of solar radiation from cloudless areas.

Increased radiative cooling of the planetary boundary layer

Aerosol particles that have grown with relative humidity become strong emitters in the thermal infrared contributing considerably to long-wave flux density divergence especially in the upper part of the boundary layer with high relative humidity values. Also atmospheric back radiation in the thermal infrared is enhanced by this effect, in parts compensating the reduction in the solar radiation range.

Stabilization of the layers with absorbing aerosol particles

If aerosol particles absorb solar radiation, this is especially so for soot, they heat the layers in which they are suspended. As most of the particles reside in the lower troposphere they warm this part of the atmosphere and less absorption takes place at the surface, because the part absorbed and backscattered by the aerosol particles no longer reaches the surface. Hence the lower atmosphere's stratification is increased suppressing or delaying convection and thus mixing of the lower troposphere, which in turn increases air pollution levels of the lower atmosphere.

At present most general atmospheric circulation models used as modules of climate models at least contain parts of the effects of aerosol particles described so far. What they generally lack is the transformation of aerosol particle size distributions through cloud processes. Due to the short life-time of a fair weather cumulus of only about 15 minutes and coalescence of cloud droplets already in such small clouds, aerosol particle size distributions are shifted to larger particle sizes, which makes them better suited for activation during the next cloud formation and increases the probability for removal by rain-out.

The next subsection will describe some of the impacts of aerosol particles on the optical properties of clouds and their life-time but will exclude largely their effects on precipitation distribution, an emerging field of atmospheric research.

2.6.2 Indirect aerosol particle effects

The existence of aerosol particles even in the most pristine polar air or in free tropospheric air guarantee cloud droplet formation at very low super-saturation in water clouds. Even in strong updrafts within convective clouds super-saturation will not surmount a few percent. However, the stronger the updrafts and the lower the aerosol particle number density (particles per unit volume) the higher is the percentage of the particles activated as cloud condensation nuclei, if they are not hydrophobic or too small for activation at the super-saturation reached. A typical minimum radius of activation is 0.02 µm. Cloud droplet concentration and size distribution is therefore not only a function of updraft but also of aerosol particle size distribution, their chemical composition and – mainly for stratus clouds – of the radiative cooling rate in the atmosphere and after cloud formation of the top layers of a cloud.

Many different influences on cloud properties are thus due to aerosol particle characteristics. In this section only the semi-direct, the first and second indirect effect of aerosol particles and the lifting of clouds by air pollution will be discussed. At the end a potential physical mechanism of cloud bursts driven by air pollution will be presented.

The semi-direct effect of aerosol particles

Cloud layers sometimes disappear during the day, because absorption of solar radiation by organic aerosol particles including soot within and above the clouds warm the cloud layer. This effect has been observed in the large continental-scale air pollution plume over India and the adjacent Indian Ocean by Ramanathan et al. (2005) during the Indian Ocean Experiment (INDOEX). It has been termed semi-direct aerosol effect as the result is a direct radiative forcing by aerosol particles which is driven by the absorption potential of particles. This adds to the tendency for warming in contrast to non-absorbing or only weakly absorbing aerosol particles.

The first indirect aerosol effect

Already Twomey (1974) and Grassl (1975) published numerical studies of optical cloud property changes caused by either more aerosol particles or high soot content of an aerosol particle population. In the first case cloud albedo increases with aerosol particle number for thin and thick clouds in the second it decreases. Depending on the optical depth or geometrical thickness of clouds, the addition of particles that also contain some soot will make them look either brighter (thin clouds) or darker (thick clouds), if looked from above. The overall effect for a region is thus depending on the distribution of cloud optical depth. In other words: thin low level water clouds will be brighter in a polluted environment while thick water clouds will become darker, but only if the relative soot content is comparably high.

Has the effect been confirmed by observations? Yes, in local studies (Raes et al., 2000) and in long satellite time series (Krueger and Grassl, 2002, 2004) for Europe and later for China, because drastic changes in air pollution after the collapse of the East Block and strong pollution increase from the 1980s to the late 1990s in China allowed a differentiation.

The second indirect aerosol effect

When air pollution by higher aerosol particle density leads to more, but smaller cloud droplets at nearly the same liquid water content the probability for coalescence of the larger cloud droplets with the smaller ones, initiating drizzle formation, is lowered. Therefore, water clouds forming in polluted air will have higher liquid water content because drizzle formation is inhibited, and thus will exist longer. This aerosol particle effect has also been observed (Albrecht, 1989) and is often called cloud lifetime effect. Its relevance on global scale is not yet assessed.

Lifting of clouds by air pollution

When water clouds form in polluted air their liquid water content stays comparably high because water removal from the clouds by drizzle is inhibited. If such an air parcel is lifted to a level where some of the insoluble aerosol particles within droplets or outside act as freezing nuclei the cloud gets glaciated and more heat is released during this phase change in polluted areas pushing the clouds higher up and thus lowering their top temperature. The drop in cloud top temperature at higher pollution levels has for the first time been observed in satellite data over Europe by Devasthale (2005) and Devasthale et al. (2005) and has recently also been observed over high pollution along shipping channels in the North Sea and the Mediterranean Sea by Devasthale et al. (2006). Lifted cloud tops radiate less to space; hence they would increase the greenhouse effect. With this new finding the overall influence of pollution by aerosol particles has to be assessed anew.

Cloud bursts caused by air pollution

Clouds freezing at higher liquid water content create larger updrafts, thus higher cloud tops and hence more intense precipitation. As cloud formation is inhibited because of lower solar radiation flux densities at the surface, reducing the height of the convective planetary boundary layer, these precipitation events are less frequent under air pollution conditions, but when they occur are more intense. This hypothesis has still to be substantiated. It is mentioned here in order to show that atmospheric physics still has many problems to solve.

2.7 Radiative Forcing of Climate Change

If radiatively active constituents of the atmosphere or the surface properties change, it must have consequences for the radiation budget of our planet, which will no longer be balanced between absorbed solar and emitted terrestrial radiation. Such a radiative balance is a prerequisite for the absence of climate change. Therefore, a measure of a disturbance is needed that easily characterizes the impact on global climate. To find such a measure will be easiest for very long-lived gases that are well mixed in the atmosphere up to about 80 km and whose absorption characteristics are well known. This is the case for all three naturally occurring long-lived greenhouse gases carbon dioxide (CO_2), nitrous oxide (N_2O), and (still acceptable as long-lived) methane (CH_4).

The following measure has been adopted by the scientific community: Radiative forcing is the change of net (solar and thermal infrared) flux density at the tropopause level calculated by fixing atmospheric composition and thermal structure of the troposphere except for the substance change in question. This so-called instantaneous radiative forcing is a fictitious value, because in reality any radiation budget change will cause climate change and thus will lead to a rearrangement of the thermal structure and the composition of especially the short-lived gases. This, however, needs time (decades) in the troposphere because of the thermal inertia of the ocean and the close connection of the troposphere to the surface, while the stratosphere can reach radiative equilibrium within some months. Despite this "fiction" the concept of radiative forcing allows an easy comparison between the climate change potential of different gases as long as their lifetime is not shorter than years. To apply it for aerosol particles or short lived gases is questionable as these particles or gases show strong regional variation, just because of their short life-time of days to weeks. Global averaging may not catch the climate change potential correctly.

Notwithstanding, the radiative forcing diagrams of the Intergovernmental Panel on Climate Change (IPCC, 2001a, 2007a) became **the** justification for climate policy measures, whether climate change is due to long-lived gases or short-lived substances in the atmosphere. Repeated here from the Fourth Assessment Report (FAR) as figure 2.4 it gives at a glance the following message:

1. Long-lived greenhouse gases (first column) show with 2.5 Wm^{-2} an accumulated radiative forcing since industrialisation began already beyond a one percent increase in solar flux density (~ 2.36 Wm^{-2}). CO_2 is dominant.
2. The climate forcing by photochemical smog (tropospheric ozone) has reached the same level as accumulated methane.
3. Several aerosol particle influences exist, which on average mask the enhanced greenhouse effect. Lowering atmospheric turbidity alone without reducing greenhouse gas emissions unmasks hidden mean global warming.
4. Land use change is only a secondary global climate change agent.
5. Aerosol research needs a major push.

Physical Climate Processes and Feedbacks

6. Solar radiation flux density change contributed only slightly to global warming in the 20th century.

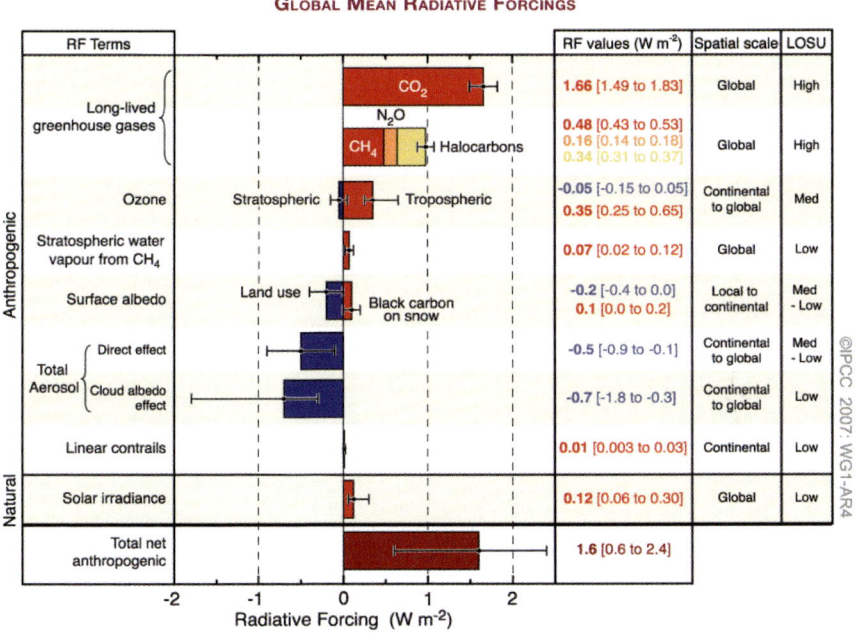

Figure 2.4   Radiative forcing of climate by human activities and changed solar radiation, taken from IPCC (2007a).

2.8 Physical Climate Processes and Feedbacks

The climate system is characterized by numerous feedbacks that involve physical, chemical and biological processes. Many of these processes are still not (well) understood. For example: Why is the carbon dioxide content in the atmosphere alternating between about 280 parts per million by volume (ppmv) in interglacials and about 190 ppmv during the coldest period of glacials (glacial maximum), although the terrestrial biosphere shrinks strongly during glacials?

Here dominant physical processes and related feedbacks will be described (in parts they have already appeared in sections 2.3 to 2.7).

2.8.1 Solar and terrestrial radiation

Whenever the solar radiation reaching the surface increases, the warmer surface will emit more terrestrial (heat) radiation and counteract further warming. In a very simple radiation balance model this can be expressed by the energy balance of a sphere with radius R

$$\text{Emission} = \text{Absorption}$$
$$4\pi R^2 \sigma T_B^4 = \pi R^2 S_o (1-\alpha)$$

$$T_B = \sqrt[4]{\frac{S_o(1-\alpha)}{4\sigma}} \qquad (2.1)$$

The average temperature T_B of the planet treated as a blackbody radiator, emitting a flux density F according to Stefan-Boltzmann's law with $F = \sigma T^4$, is only determined by solar flux density, S_o, reaching the Earth, often called solar constant or solar irradiance, diminished by the Earth's albedo α. Using $S_o = 1367$ Wm^{-2}, $\alpha = 0.3$ and $\sigma = 5.67 \cdot 10^{-8}$ Wm^{-2} K^{-4}, we get a temperature T_B of 255 K. Thus the radiation to space originating from the Earth is equivalent to a blackbody with a temperature of about -18°C. This temperature occurs in the atmosphere on average at about 5 km height, thus indicating that emission to space originates largely in the atmosphere and stems less from the surface, whose average temperature is 288 K. Hence the greenhouse effect of the atmosphere is about 33 K.

Increasing S_o by one percent leads to $\Delta T_B \approx + 1$ K. The interplay between solar and terrestrial radiation is a strong constraint for the planetary radiation budget. It also becomes clear from equation 2.1 that the climate system could counteract solar radiation changes by changes in planetary albedo. As the latter is to a large degree determined by the clouds in the atmosphere a potentially strong negative feedback is less cloud reflection in glacials and more in interglacials or warmer climate periods without any continent-wide glaciation. Whether this is true we do not know yet.

2.8.2 Clouds

Liquid cloud droplets and solid ice-crystals in ice clouds (cirrus) are minor constituents of the atmosphere with major impact. A water cloud with 500 m thickness, a typical stratus deck, reaches an optical depth δ of about 25 in the solar spectral range, although the liquid water content is only 0.2 g m^{-3} or 100 g m^{-2} liquid water column content, which is equivalent to a 0.1 mm water layer. Finely dispersed liquid water in air together with very low absorption of liquid water in the visible creates clouds as effective backscatterers that can reach albedo values up to 0.8, nearly as brilliantly "reflecting" as fresh powder snow.

Calculating spectral optical depth δ_λ of a 2 km thick water cloud extending from height z_1 to height z_2 gives

$$\delta_\lambda = \int_{z_1}^{z_2}\int_{r_1}^{r_2} Q_{ext,\lambda}^{(r)} \pi r^2 N(r,z)dr\,dz \approx 100 \quad (in\,the\,visible\,spectral\,range) \qquad (2.2)$$

We realize that the cloud's impact on the radiation budget is mainly a function of its droplet size distribution $N(r, z)$, which also varies with height z and not only with droplet radius r, but is also proportional to the cross section πr^2 of the droplet and the spectral extinction efficiency $Q_{est,\lambda}$ of a Mie-scatterer. $Q_{est,\lambda}$ is also close to 2 for cloud droplets in most parts of the solar radiation range with $\lambda \leq 2\mu m$. Direct solar radiation transmission of many clouds $T_r = e^{-\delta_\lambda} \approx 0$, i.e. we cannot see the sun's disk. As our eye can distinguish radiance L_λ differences of about 2 percent the sun's disk disappears for our eye at $\delta \approx 10$, when $L_\lambda(sun)\,e^{-10} \approx L_\lambda(sky)$.

In the thermal infrared (terrestrial) radiation range at wavelength > 4 μm liquid water is a strong absorber, hence already thin clouds with δ > 3 (50 m fog layer) fully decouple long-wave radiation transfer above and below clouds, while we can still read a newspaper under a cloud with $\delta \approx 100$ at high solar elevation because very low absorption by liquid water allows scattered solar radiation to reach the ground.

Therefore the following feedbacks of clouds exist (as also seen in table 2.4):

1. High, optically thin ice clouds enhance the greenhouse effect of the atmosphere, most contrails from airplanes belong to this category.
2. Low, optically thick clouds counteract the greenhouse effect of the atmosphere; especially stratocumulus decks off the western subtropical coasts of the American and African continents belong to this category.
3. Many clouds will exist neither enhancing nor reducing the greenhouse effect; the sign of their feedback will depend on subtle changes of cloud height and thus temperature, mean droplet radius, crystal shape, geometrical extent, three-dimensional structure and even surface albedo and – of course – solar zenith angle.

Table 2.4: Cloud cover change and its feedback on the greenhouse effect

Cloud type	Feedback sign and strength	Key influencing factors, attempt to rank
stratus [3] stratocumulus ground fog	- strong - strong - very strong	shortwave albedo small temperature difference between cloud top and surface
fair weather cumulus	- medium - medium	cloud cover height of top
cumulus congestus	- medium	cloud cover cloud top temperature
altocumulus	- or +	cloud top temperature optical depth cloud cover
altostratus	- or +	cloud top temperature optical depth
thin cirrus	+	optical depth cloud top temperature
cirrostratus cumulonimbus	+, rarely - ?	optical depth cloud top temperature vertical extent

Overall, clouds lead to a reduced greenhouse effect of the atmosphere at present, compared to the cloudless regions. Their effect on the radiation budget amounts to about -15 Wm^{-2}, which is equivalent to a planetary albedo change of about 6 percent. Whether they loose or gain in their cooling capacity when a further surface warming occurs is not clear yet. To find out what will happen to their distribution, lifetime, microphysical properties, etc. is a key research question.

2.8.3 Intensity of the meridional overturning in the Atlantic

From reconstructions of climate history in the North Atlantic region it became clear that its thermohaline circulation (driven by density structure, which in turn is due to vertical temperature and salinity structure) has changed frequently during glacials. Our present interglacial, the holocene, is largely defined as the period after the last stop of strong meridional overturning in the North Atlantic during the so-called Younger Dryas 12,000 to 11,500 years ago. At present we observe an intermittent strong convection in waters close to Greenland either in the Greenland Sea or the Labrador Sea. There, mostly at

---

[3] More stratus would strongly cool the Earth's surface.

the end of winter, water density at the surface is high enough for convection to take place down to sometimes more than 2 km creating North Atlantic Deep Water (NADW) that then circulates in the global ocean as part of the global conveyor belt (see figure 2.5). The physical background is that rather saline Atlantic water is reaching high northern latitudes where it is cooled close to or to the freezing level of sea water (-1.8 to -1.9 °C, depending on salinity). About 4 to 6 million m^3 per second (10^6 m^3 s^{-1} are called 1 Sverdrup by oceanographers) are thus sinking into the ocean interior (e.g. the Greenland Sea).

Flowing over sills between Greenland and Iceland as well as between Iceland and Scotland, these water masses entrain more "Sverdrups" so that finally about 15-20 Sv flow southward in several kilometers depth mainly along the North American continental slope. As a reaction to this vertical portion of the global conveyor belt warm near surface waters flow northward in the eastern North Atlantic off Europe, creating Europe's mostly mild climate with about 10^{15} Watt given to the atmosphere. Falsely named the Gulf Stream, correctly the North Atlantic Drift, this ocean current (discontinuous, often in form of meanders and drifting eddies) makes Europe so warm in comparison to similar latitudes in the Eastern Pacific.

Will the thermohaline Atlantic overturning circulation weaken or even stop in a warming world? This potential threat has got intense media coverage and present knowledge can be summarized as follows:

1. Coupled atmosphere/ocean-models running under enhanced greenhouse effect conditions show a decrease or even a complete stop of the thermohalinen overturning circulation in the northern North Atlantic depending on the strength of climate change (see figure 2.6).
2. The few observations of the meridional overturning circulation in sections across the North Atlantic at about 27°N do not allow a firm statement about a recent weakening, as natural variability on decadal time-scales is not well known.
3. In a strongly warming world North-west Europe might become a region with reduced warming in the 21st century due to the weakening of the North Atlantic drift, as a consequence of less meridional overturning.

The thermohaline circulation would accordingly feed back negatively to the enhanced greenhouse effect, but with effects of mainly regional character. Nevertheless this could mean a major threat by extreme weather as stronger gradients of temperature would occur in the region affected.

2.8.4 Shift of extratropical storm tracks

The climate in mid-latitudes as well as in higher latitudes is strongly depending on tracks and intensity of mid-latitude cyclones. The stimulus for their formation are meridional temperature gradients in the free troposphere. Their intensity depends in addition on the amount of humidity in the atmosphere. The statistics of cyclones on global scale has revealed that the North Atlantic region experiences winds as intense in winter as within the circumpolar Southern Ocean storm track, where strong winds prevail nearly throughout the entire year.

Will the storm tracks change or have they already changed?

Evaluations of long time series of pressure measurements at several stations, which are more reliable for pressure gradient estimates and thus geostrophic wind vectors than direct wind vector measurements, have shown that there is no general intensification of gale force winds. However, there is an indication of a northward shift of cyclone tracks in the Atlantic. Model results for climate change scenarios in the 21st century point to a clear northward shift jointly with an intensification mainly in the southern part of the storm track, i.e. for Great Britain and Northern Central Europe. The plausible explanation is: Despite the northward shift, a consequence of reduced meridional gradients, intensification occurs due to enhanced latent heat release especially in the southern portions of the cyclones as the air is warmer and contains more humidity.

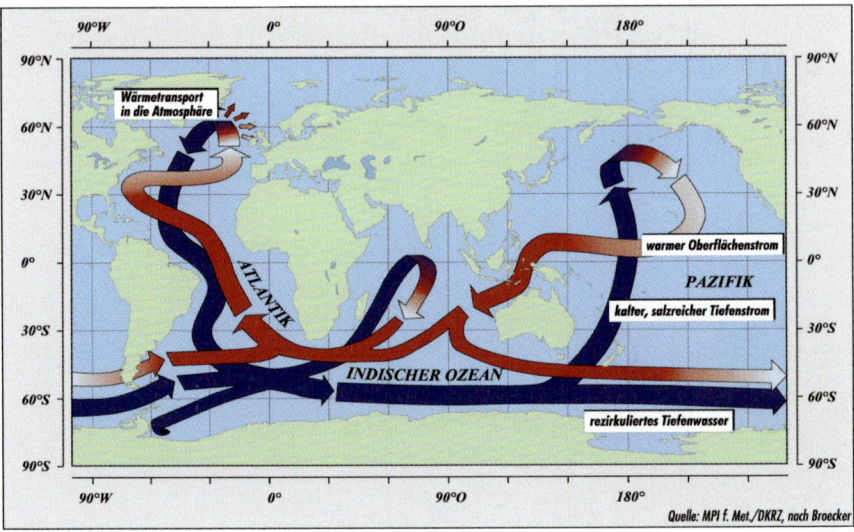

Figure 2.5: **The global conveyor belt in the world ocean**

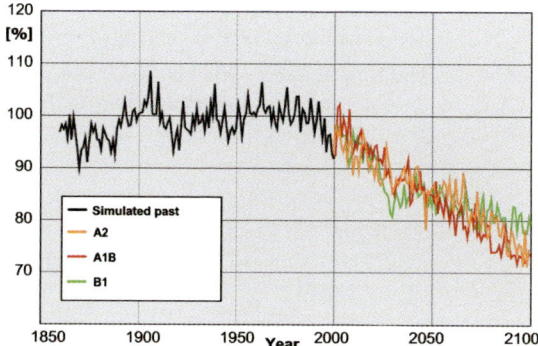

Figure 2.6: Weakening of the thermohaline overturning circulation as a function of climate change scenarios (MPI, 2006b).

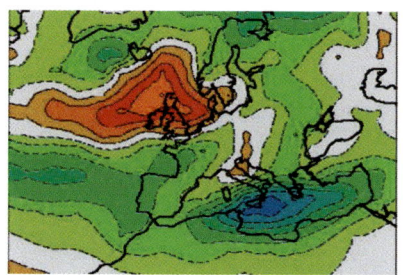

Figure 2.7: Changes in the number of cyclones per month in the northern North Atlantic for the winter period (DJF) in the climate change scenario A1B (Bengtsson, 2006)

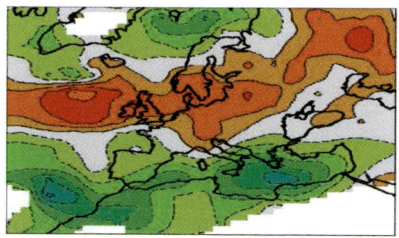

Figure 2.8: As figure 2.7 but for storm intensity

2.8.5 Soot versus cloud condensation nuclei

If aerosol particles are more numerous and also soluble, for example consist of ammoniumsulfate ((NH_4)$_2$ SO_4) formed from the air pollution gases sulphur dioxide (SO_2) and ammonia (NH_3), clouds would contain more and smaller droplets per unit volume at the same circulation conditions (see also section 2.6). These clouds scatter more sun-light (see equation 2.1) and their albedo would increase, counteracting an enhanced greenhouse effect, if their height would not change. If polluted air – as is normally the case – would also contain more soot (black carbon) the clouds forming in such air would be less scattering, look darker when viewed from above and below. Only for optically not very thick clouds would the increased light scattering effect dominate under such conditions. Hence, soot will damp the clouds' ability to mask effects of an enhanced greenhouse effect. Therefore addition of soot would contribute in the cloud free atmosphere to the warming but dampen the indirect effect of air pollution on cloud albedo, certainly overriding it for optically thick clouds.

Reducing soot content of the air, a health issue anyway, has also a potential to reduce global warming.

2.9 Atmospheric Chemistry and Climate

Many chemical reactions in the atmosphere are climatically relevant because they change concentrations of radiatively active gases or particles as well as could properties. Two global phenomena of this kind are at least partly understood and one of these has led to the first successful global environmental protection policy. Both phenomena are part of the atmospheric budget of greenhouse gas No. 3, namely ozone. The first phenomenon is depletion of stratospheric ozone by chlorine and bromine containing compounds derived from halogenated hydrocarbons, the second is ozone formation in the troposphere driven by air polluting precursor gases in the sun-lit atmosphere.

2.9.1 Stratospheric ozone depletion

It was a full surprise, not imagined by the scientific community, when strong reductions of ozone column content observed over two stations in the Antarctic, a Japanese and a British one, were published in the mid 1980s. Soon thereafter balloon soundings and the evaluation of archived satellite data made it clear that in Antarctic spring ozone was nearly completely absent in stratospheric layers from 12 to 20 km height that normally contain highest ozone concentrations. Therefore many speak of an ozone hole. Soon the cause became known as well. When sunlight comes back after polar night chlorine and bromine containing compounds are transformed at the surface of polar stratospheric cloud particles, composed of nitric and sulphuric acid plus water – existing only at temperatures below -78°C – into new chemicals that are part of a catalytic ozone destruction cycle. The overwhelming part of the chlorine and bromine containing compounds stems from halogenated hydrocarbons used by mankind for many different purposes like cooling, cleaning, foaming and spraying as well as extinguishing fires. Three air chemists, Crutzen, Rowland and Molina got the Nobel Prize in chemistry in 1995, because of their fundamental work in the 1970s when these compounds and catalytic reaction chains got into a first environmental protection policy debate. In the meantime the *Montreal Protocol* of 1987 as part of the *Vienna Convention to Protect the Ozone Layer* of 1985 had been enforced several times and chlorofluorocarbons and other compounds had been phased out in industrialized countries, the main producers and consumers.

Many of the ozone depleting substances have started to decline in the beginning of this century. But because of the long lifetime of these compounds and concomitant effects caused by the enhanced greenhouse effect in the stratosphere, recovery of the ozone layer in the stratosphere will need at least several decades, if no other substances attack this life-supporting layer of a minor atmospheric constituent.

2.9.2 Photochemical smog or increased tropospheric ozone

The IPCC Fourth Assessment Report (IPCC, 2007a) shows that the radiative forcing of climate by tropospheric ozone has reached the same level as accumulated methane with about 0.3 Wm^{-2}. Ozone in the troposphere is formed by a totally different chemical reaction chain than in the stratosphere where solar radiation at wavelengths < 0.24 μm dissociates oxygen molecules into two oxygen atoms, which can combine with other oxygen molecules to form ozone (O_3). In the troposphere so-called precursor gases, namely nitrogen oxides (NO_x = NO + NO_2) and hydrocarbons are needed above a certain concentration level of NO_x to form ozone. The key reaction determining the ozone formation rate is the dissociation of nitrogen dioxide (NO_2) into nitrogen monoxide (NO) and an oxygen atom (O) by ultraviolet radiation at wavelengths λ < 0.4 μm, which can penetrate the atmosphere down to the surface. As large quantities of hydrocarbons are also emitted by vegetation the key pollution gas in this context is nitrogen dioxide originating from traffic, power plants, heating of buildings by oil and gas and industrial processes but also from vegetation fires, especially deforestation by slash and burn practices. Tropospheric ozone is therefore a continent wide air pollution problem. Because ozone formation rate reaches a maximum at a certain NO_x to hydrocarbon ratio, it is difficult to forecast effects of clean air acts as reduction of one component, e.g. NO_2, might cause higher ozone levels far away from the emission source.

Photochemical smog, as it is often called, has become a nearly global phenomenon and its abatement is much more difficult than anticipated. The involvement of so many environmental factors makes forecasting of effects of air pollution control a real challenge. Whenever authorities report near surface ozone concentrations above the alarm level (180 µg m^{-3} O_3 in ambient air in several European countries) this has also a climate change importance.

Many more chemical reactions among trace gases as well as between trace gases and aerosol particles are relevant for climate but are not discussed here. However, they do not have the same importance as the changes of ozone. While we reduced it where we need it as a UV radiation shield (in the stratosphere) we augmented it by more than a factor of 2 in the 20th century in the troposphere where it is an air pollution component.

The overall climate effect of the changed ozone profile is difficult to assess because we have to include halogenated hydrocarbons, very potent greenhouse gases, into the discussion, too. In addition, even no change in the total ozone column content as a result of depletion above and augmentation below the tropopause can have a strong climate effect, as we redistribute the solar radiation absorption thus heating rates in the vertical and also the vertical profile of terrestrial radiation flux density is changed, leading to changed cooling rates.

2.10 Climate Change and Vegetation

Vegetation largely determines or at least influences greenhouse gas concentrations, lowers surface albedo and increases evapotranspiration, dominates surface roughness and reduces run-off. Therefore it is a decisive component of the climate system. Often vegetation is used as a term including microbes, fungi and animals living in it, although the term terrestrial biosphere would be more appropriate then.

In the context of climate change vegetation became a hot topic because very different views on its role in climate change policy exist. On one hand some see reforestation as a key strategy to reduce the anthropogenic CO_2 burden in the atmosphere, on the other hand vegetation is described as a presently growing carbon reservoir which might turn into a CO_2 source above a certain climate change threshold, thus amplifying mean global warming in the latter part of the 21st century.

In order to grasp the complexity of the issue let us consider reforestation for two different forest types, first the boreal and then the tropical rain forest.

2.10.1 Climate impact of reforestation in the boreal zone

Several climate parameter shifts will occur:

Albedo change: Strong reduction of surface albedo from about 80 percent in late winter to about 20 to 30 percent if the forest is regrown. General decrease in all seasons.

Evapotranspiration change: The forest will increase evapotranspiration compared to grassland especially within the growth period.

Carbon storage: The carbon stored additionally in existing old forests per hectare per year depends on soil type, climatic zone and tree species. Numbers have reached about 1 tC/(ha•a).

Trace gas emissions: Emissions of N_2O and CH_4 could increase if no artificial fertilization of the grassland has taken place earlier. Higher emissions of non-methane hydrocarbons.

Emissivity in the thermal infrared: All forests show reduced emissivity because they approach a blackbody radiator better than grassland.

Does the warming caused by the strong albedo increase that is only in parts reduced by increased evapotranspiration override the reduced greenhouse effect due to forest regrowth? In other words: Would reforestation in the boreal climate zone lead to further global warming?

The final answer can only be given by coupled model calculations. A back-of-an-envelope calculation gives a yes to the above questions (see also chapter 3 on climate modelling).

2.10.2 Climate impact of reforestation or deforestation in the tropics

In contrast to the boreal forest zone reforestation in the tropics has strongly different impacts on regional and global climate. The albedo decrease is modest, about five percent if grassland is converted into a forest; evapotranspiration increases strongly, off-setting or surmounting the albedo influence; carbon storage is much stronger within the first decades; emissions of N_2O and CH_4 from tropical forest soils are on average larger than from the soils in the boreal zone. Again only coupled model calculations can give an answer. However, the parameterization of evapotranspiration from forests as compared to grassland in inhomogeneous terrain is still a research topic, i.e. error bars are still high. It is, therefore, very difficult to assess the climate change caused by deforestation in the Amazon basin and model results diverge strongly.

3 Climate Modelling

If the basic laws governing the physics and the chemistry as well as empirical relations between biospheric, chemical and physical parameters are known, numerical models of the Earth system can be developed that allow to estimate potential future states of the system provided the external forcing (e.g. anthropogenic forcing) can be estimated. If only the physical processes, the external radiative forcing, e.g. by greenhouse gas concentration changes, but no interactive biogeochemical cycles except the water cycle (the most important one) are included, we speak of climate models.

3.1 Model Basics and Structure

These models solve coupled, prognostic, non-linear, partial differential equations on a three-dimensional grid for both the atmosphere and the ocean. The coupled equations are the consequence of the following physical laws: Newton's second law, mass conservation, first and second law of thermodynamics, equations of state for air and ocean water, budgets of air humidity, liquid water in the atmosphere, in soils and rivers as well as cloud ice and snow, radiation laws of Planck and Kirchhoff, radiative transfer equation, spectral absorption by gases involving some quantum mechanics (see also box 3.1). In addition, empirical relations between vegetation, bare surfaces and radiation parameters are needed. Another complication in solving these equations arises from the changed surface characteristics during the time integration, e.g. due to snowfall or snowmelt. Many physical processes have to be parameterized, because the spatial resolution of the models cannot resolve the manifold sub-grid scale processes like the influence of small clouds or even cloud ensembles. Such parameterizations are often derived from dedicated field experiments.

Box 3.1: **Equations and Empirical Relations in a Climate Model**

All the equations shown are simplified ones as the full equations would, for example, also contain sound waves, whose high speed would force the numerical scheme to extremely small time steps for the solution of the prognostic equations.

All these equations relate the key variables like velocity vector $\vec{v}$, pressure p, acceleration of gravity $\vec{g}$, rotational speed of the Earth $\vec{\Omega}$, temperature T, density ρ and absolute humidity with diabatic (i.e. entropy generating) processes (in rectangles) originating from friction at the surface $F_{\vec{v}}$, net radiation flux divergence Q, surface heat flux F_T, phase fluxes of water S_q and humidity fluxes at the surface.

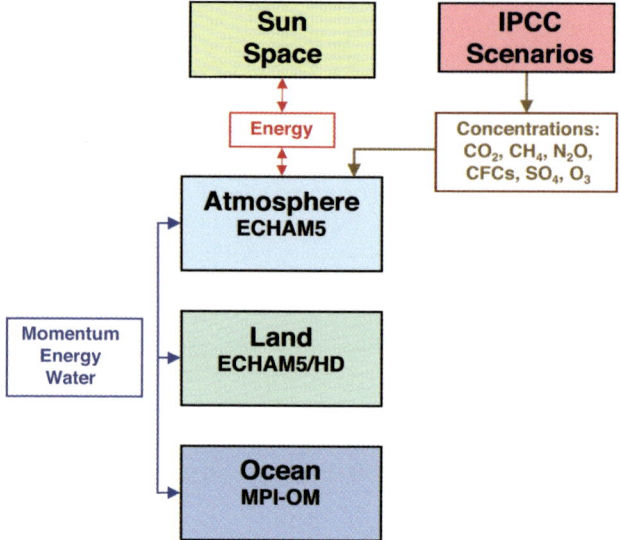

Figure 3.1   Schematic diagram of the climate model components used for typical climate change runs. The general circulation model of the atmosphere also includes land surface processes like evapotranspiration, river run-off and the ocean module contains sea ice thermodynamics and rheology. Volcanism as well as solar radiation changes is handled as external processes like atmospheric trace substance concentration changes due to anthropogenic activities (MPI, 2006b)

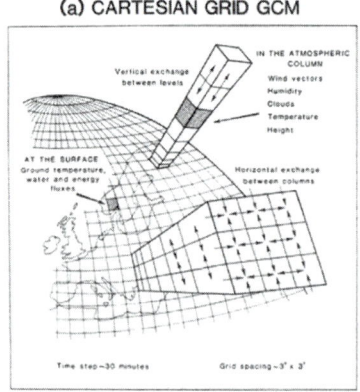

Figure 3.2   Schematic representation of the 3-dimensional grid of a climate model

3.2 Climate Model Evaluation

Over the last decade climate model evaluation became an activity of globally co-ordinated research programmes especially the World Climate Research Programme (WCRP). A thorough evaluation helps to understand deficiencies of such coupled atmosphere/ocean/land models and thus shapes their suitability for answers to certain scientific questions.

The following steps of climate model evaluation should be successfully done before the model can be used for projections of future climate that goes well beyond climate states known from the instrumental period.

1. Reproduction of present day climate including observed variability and thus extremes.
2. Reproduction of the climate of the 20th century for which radiative forcing and global climate have been derived or measured.
3. Reproduction of the last millennium in which forcing by volcanoes and solar flux density variations could be reconstructed from ice cores, lake sediments, tree-rings, corals, etc. at least for the northern hemisphere.
4. Reproduction of an abrupt climate change event in recent climate history using palaeoclimatic evidence in so-called proxy data.

While models of all major climate research centres have successfully passed the first two tests only few have passed the third and the fourth has been attempted only with so-called Earth system models of intermediate complexity (EMICs) that strongly simplify modules for the atmosphere, the ocean and the vegetated land surface.

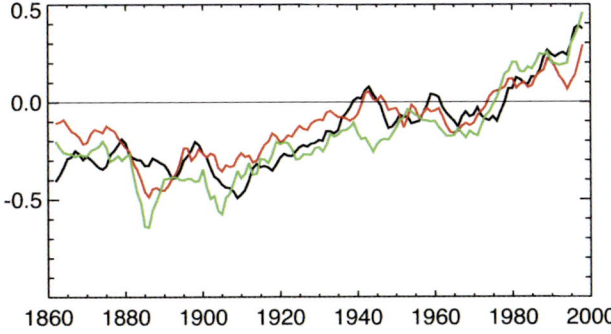

Figure 3.3   Modelled (red) and measured (black) global mean near surface air temperature since 1860. Using the interactive aerosol particle module the green curve results. Because of internal variability on inter-annual and decadal time scales the curves must not coincide but should show similar long-term trends (MPI, 2006b)

As an example for model validation the global mean temperature change since 1860 is shown as observed and modelled in figure 3.3 (MPI-M, 2006), whereby the model input also contained natural climate change factors like major volcanic eruptions and changes of solar irradiance at the top of the atmosphere. When the interactive aerosol module has been used (green curve) the temperature decrease after the major volcanic eruption of Krakatao in 1883 became too strong, probably a consequence of the insufficient knowledge about emitted SO_2.

3.3 Emission Scenarios

The future behaviour of humankind with respect to its energy supply system transformation is largely unknown. Therefore, a broad range of emission scenarios has been developed under the assumption of no climate policy going beyond the first emission reduction step, the Kyoto Protocol (IPCC, 2000). These so-called SRES-Scenarios (Special Report on Emission Scenarios) have been used again for the Fourth Assessment Report (FAR) of the Intergovernmental Panel on Climate Change (IPCC) that has been published in 2007. As shown in Table 3.1 their CO_2-emissions differ massively, ranging from 29 GtC per year in 2100 to only 4 GtC per year for scenarios A2 and B1, respectively. While scenario A2 assumes a world with strongly differing regional development, continued very high population increase in developing countries and low environmental consciousness, scenario B1 assumes global economic development (convergence), hence with reduced population growth at high environmental consciousness (but still no stringent climate policy) and technological development. Scenario A1B assumes strong economic growth, an energy supply system that has balanced (B) contributions from fossil and renewable energy resources, and a population peaking at 9 billion heads in 2050. For more details on other sub-scenarios of A and B please consult IPCC (2000). The atmospheric CO_2 concentration reaches 540 ppmv for B1, 700 ppmv for A1B and 830 ppmv for A2 in 2100.

B1* describes a world with the same population as in the family of scenarios A1 but a more rapid transformation into an information society with lower material fluxes and low emission technologies.

Table 3.1: CO_2 and SO_2 emissions in the 21st century according to three different scenarios (GtC/a and MtS/a, respectively)

	CO_2			SO_2		
Year	A2	A1B	B1*	A2	A1B	B1
2000	8	8	8	69	69	69
2020	12	13	11	100	100	75

	CO_2			SO_2		
2040	16	15	12	109	69	79
2060	19	16	10	90	47	56
2080	23	15	7	65	31	36
2100	29	13	4	60	28	25

As shown in figure 3.1 climate models need prescribed trace substance concentrations, typically methane (CH_4), nitrous oxide (N_2O), chlorofluorocarbons (CFC), ozone (O_3) and sulfate particles (SO_4) are taken into account. Some recent model configurations have interactive aerosol modules, i.e., they do not prescribe SO_4 but accept SO_2 emissions and some also can handle soot (black carbon).

3.4 Projections of Climate Change

Since greenhouse gas emissions are uncertain we cannot get climate change forecasts or predictions. We can only project climate change for a given emission scenario as a plausible future. Many climate research centres have used the above and other emission scenarios and have projected future climate, typically until 2100 but some are also going beyond that date. All these projections have been assessed by IPCC. Here results from one of the contributing centres, the Max Planck Institute for Meteorology in Hamburg, Germany, will be presented. As all the scenarios described and prescribed for the model did not contain a dedicated climate protection policy, the results cannot contain a feedback between climate change realized at a certain time and the emissions thereafter, be it due to economic pressure or through planned emission management. Hence, the scenario results are mere input for climate policy development within countries or the global community under the UNFCCC umbrella.

Figure 3.4 shows a "shocking" result: The 21st century would see a global mean temperature change of 3 to 4 K, nearly the same as the temperature difference between the last glacial maximum (18,000 years ago) and the present interglacial (Holocene), which was 4 to 5 K, but now pressed into one century. Only scenario B1 stays well below with about 2.5 K warming until 2100. We have clearly entered the anthropocene. Further inspection of figure 3.4 points to extreme warming in the Arctic and lowest warming in areas with a deep mixing ocean (Southern Ocean, northern North Atlantic). Continents warm generally stronger than ocean areas, except the Arctic Ocean where sea ice loss leads to more than double the average warming.

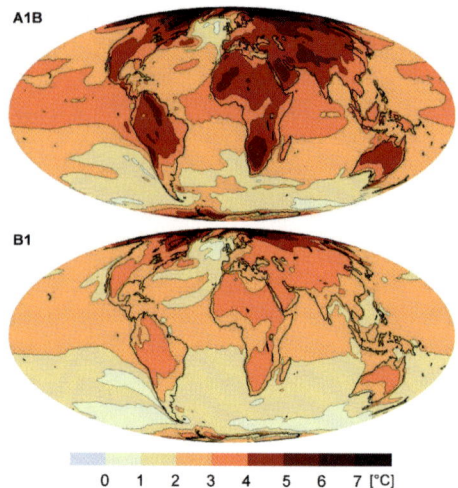

Figure 3.4 Near surface air temperature change in the 21st century for scenarios A1B and B1. The values were calculated by comparing the 30 year period from 1961-1990 with the 2071-2100 period (MPI, 2006b)

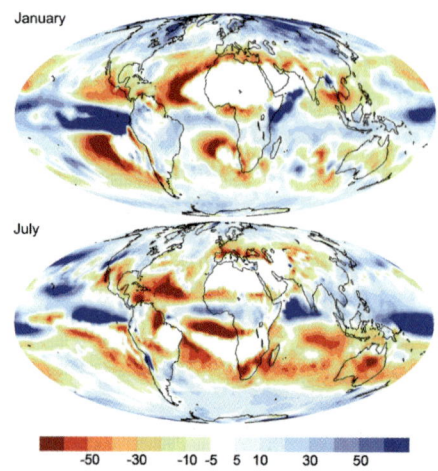

Figure 3.5 Relative precipitation changes in percent for scenario A1B in January and July (comparison between 2071-2100 and 1961-1990). Changes can surmount 50 percent in both directions.

Projections of Climate Change 35

For many areas precipitation changes are more important than temperature changes. Therefore figure 3.5 is of high interest. A general shift of rain belts emerges: more precipitation in the inner tropics and most high latitudes of both hemispheres but much less in the outer tropics and sub-tropics. Slightly exaggerated: where it rains already enough it will become more and less for those having often not enough, although overall precipitation increases by 5 (scenario B1) to 7 percent (scenarios A2 and A1B) until 2100.

How would sea level change for the above scenarios? One has to take into account three processes, namely density changes of sea water (mainly thermal expansion), melting of ice on land (glaciers and ice sheets) and changes in ocean circulation. The largest contribution in the 21st century will come from sea water density changes (+21 cm for B1, +26 and 28 cm for A1B and A2, respectively) and regionally from rearranged ocean circulation. Melting of ice on Greenland would contribute +13 cm (A1B) while for the Antarctic ice sheet the model calculates -5 cm. Figure 3.6 clearly shows the strong redistribution of sea level rise through ocean circulation changes. Sea level rise for example in the Arctic is about 0.5 m for scenario A1B due to density changes of sea water (~ + 25 cm) and intensified westerly flow (+ 20 cm) as well as melting of land ice (+ 0.08 m).

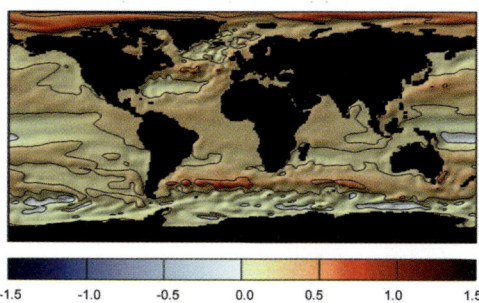

Figure 3.6 **Sea level change (mostly rise) due to sea water density changes (mainly thermal expansion) and changed atmospheric as well as oceanic circulation for scenario A1B. Please note the high values in the Arctic Ocean where less saline water leads to lower density and thus higher levels despite modest temperature changes in the ocean interior (MPI, 2006b)**

As it became clear from figure 3.4 for the air temperature changes already, the Arctic experiences the strongest warming due to the positive ice/snow albedo-temperature feedback (see section 2.5). Scenarios A1B and A2 lead to an ice free Arctic Ocean at the end of summer, which certainly would have a major impact on marine and terrestrial ecosystems there and in adjacent areas. Figure 3.7 points to these dramatic changes both for sea-ice and snow cover on land.

Snow cover in September, nowadays common in northern Siberia, Canada and Alaska will be largely gone.

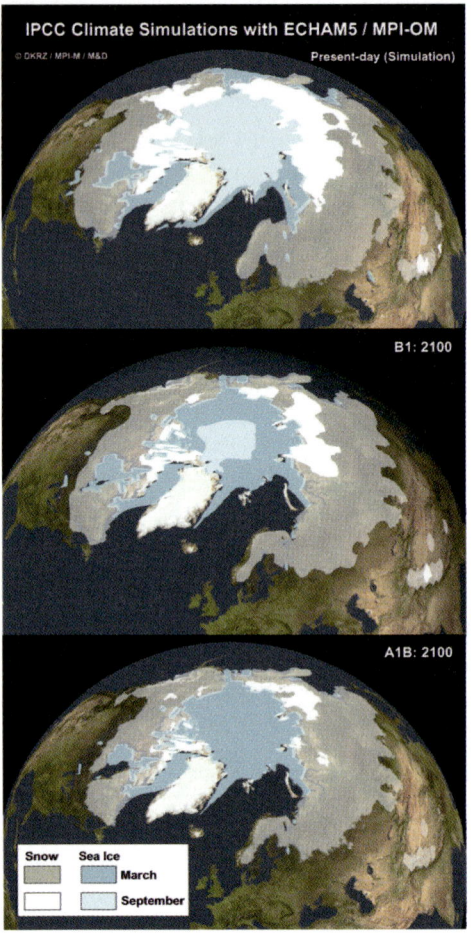

Figure 3.7 Sea ice and snow cover on land in March and September for present climate (model result as well) and for scenarios B1 and A1B in 2100

Frequently also a breakdown of the so-called thermohaline circulation in the North Atlantic is discussed and used for guard rails in climate policy development. This thermohaline circulation transports warm ocean waters from the tropics to high latitudes and cold water masses in the deep North Atlantic southward. A reduced sea water density in near surface layers of the high latitude ocean either through warming or more precipitation (both phenomena are forecast) would disturb this overturning circulation or in extreme cases stop it. What do the most comprehensive models to date project for the emission scenarios? Yes, the overturning circulation diminishes by about 30 percent until 2100 leading to less warming in parts of the North Atlantic but with a rather modest influence on Europe. Further model experiments with additional melt water fluxes from Greenland also did not lead to a breakdown. However, in model runs lasting millennia such a breakdown was simulated for the time after the year 2300 in 3 of 5 realisations of scenario A1B and always in scenario A2.

Can the model results with such a low resolution of about 200 km at the equator also point to extreme weather event changes? As we know from statistics of damages by such events, published by the Munich Re-insurance Company, the frequency of and the damages by such catastrophes have dramatically increased, however, mainly through local misbehaviour (e.g. buildings in flood-prone areas). A certain category of such extreme events can be modelled already: those happening over larger spatial and temporal scales like maximum precipitation over 5 days causing large river floods or the length of dry spells leading to droughts. The coupled model results for the 20th century show an increase in the length of heat waves, a reduction in frost days as well as an increase in maximum 5-day precipitation. Are these trends continuing? As figure 3.8 for scenario A1B underlines the 5-day maximum precipitation amount in a year increases nearly on global scale except for North Africa and some Mediterranean climate areas. In other words: the frequency and intensity of major floods will increase in most regions. It is also clear from figure 3.8 that tropical areas will experience the strongest absolute precipitation increase.

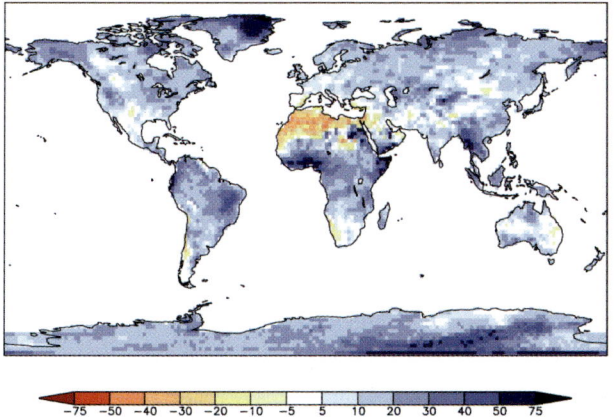

Figure 3.8 **Changes of 5-day extreme precipitation per year (in percent) for scenario A1B (2071-2100) in comparison to observed values from 1961-1990 (MPI, 2006a)**

3.5 Regional Climate Change Information

It is a well accepted rule not to interpret the results of a numerical model at space scales smaller than about four times the resolution of the numerical scheme. Therefore a global climate model with 200 km horizontal resolution can only give information at scales of about 1000 km. If climate change information at scales below 1000 km is sought regional models have to be nested into global models. As obvious from figure 3.9 even double nesting is used in recent years to project climate change to scales of about 10 km.

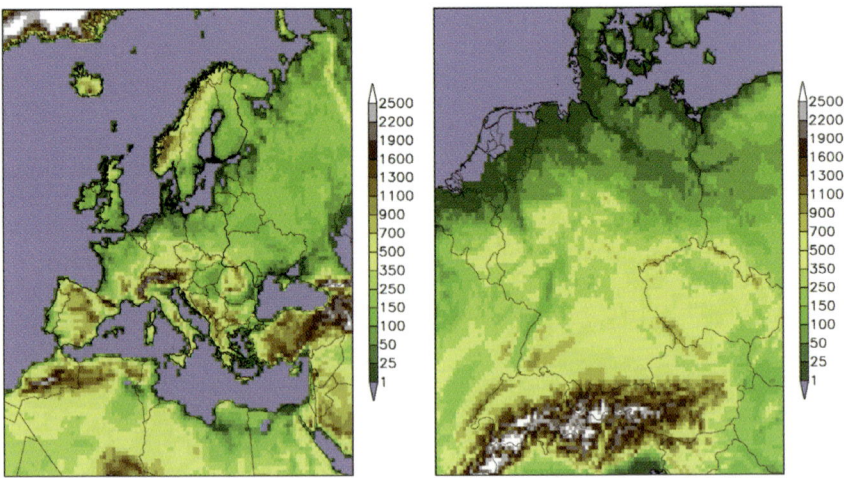

Figure 3.9: The spatial scales of regional models. This schematic demonstrates what is typical for present day regional climate models

3.5.1 Why is regional climate modelling needed?

Regional climate modelling is needed firstly, because regional climate is determined by the interaction of large-scale processes, e.g. travelling cyclones, and regional scale processes. Large scale circulation determines the statistics of weather events that characterize the climate of a certain region. But also regional and local scale forcings and regional special circulations, like valley winds, modulate the regional climate change signal, which could even feed back into the large scale.

Secondly, in order to simulate climate change at the regional scale it is necessary to simulate processes at a wide range of spatial and temporal scales. The more so if strong orographic features and land use differences exist.

Regional Climate Change Information

However, it must be stated here that systematic errors in the large scale forcing fields cannot be corrected. Hence global model biases remain at regional scales. In addition, the parameterizations of unresolved physical processes should be adapted to the different scales and feedbacks into the global scale should be taken into account. At present only first attempts to perform both tasks exist.

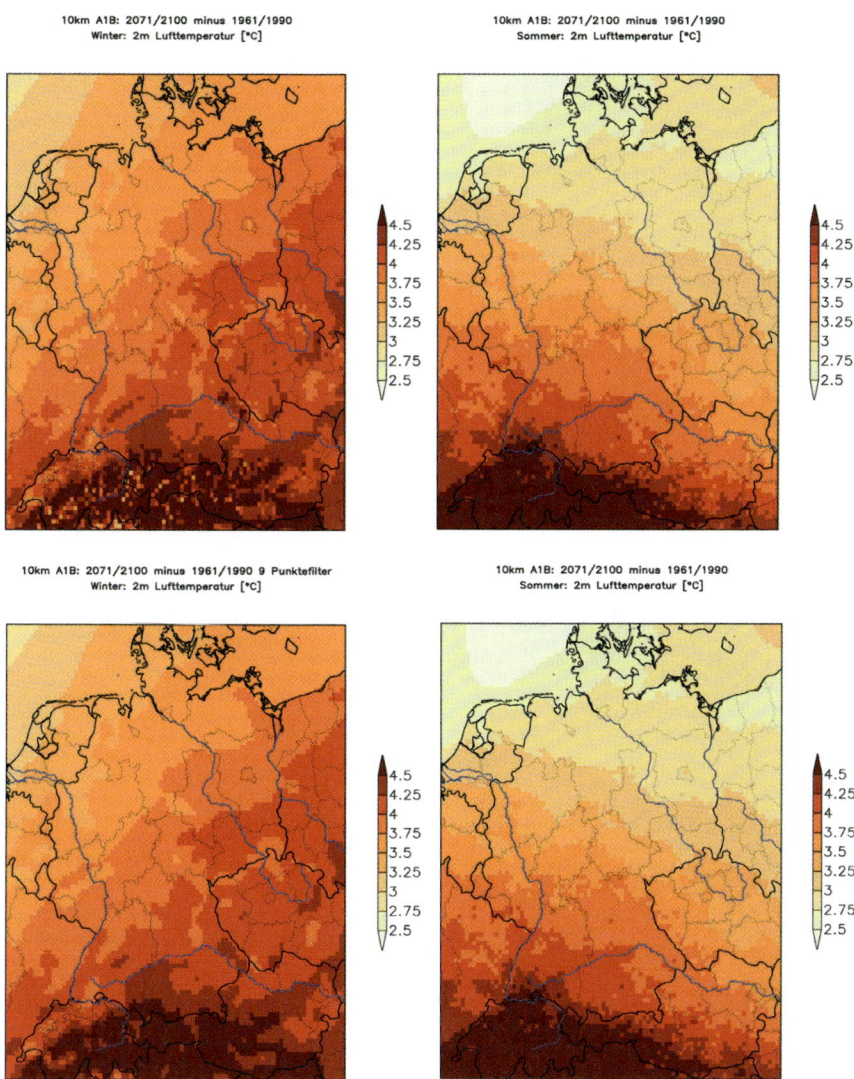

Figure 3.10: Average winter and summer temperature change at the end of the 21st century (2071-2100) in Central Europe, for the emission scenario A1B (MPI, 2006b)

The following examples are taken from a new regional modelling attempt for the IPCC scenarios at the Max Planck Institute for Meteorology in Hamburg where double nesting down to horizontal scales of 10 km has been performed for several scenarios using the global model results as lateral forcing. As

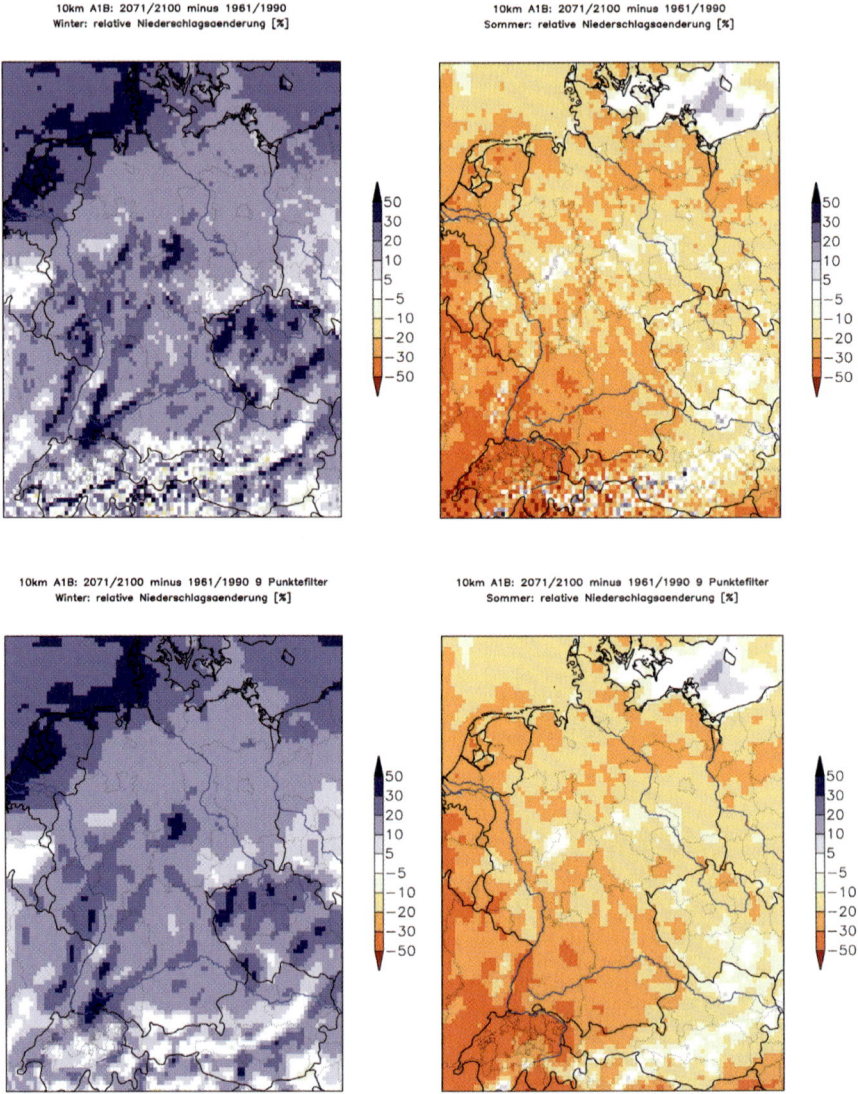

Figure 3.11   Average winter and summer precipitation change at the end of the 21st century (2071-2100) in Central Europe for the emission scenario A1B

clearly visible in figure 3.10, both central European summer and winter become much warmer at the end of the 21st century, with a strong tendency for strongest warming in the south. When looking at the precipitation change (figure 3.11) for the same period in scenario A1B winter time wetness increases as well as summer dryness. In addition, the changes are more pronounced in the hilly terrain than in the flat lands. It is clear from many more results (not shown here) for days with snow or frost, highest precipitation events, etc. that many new weather extremes would result and known extremes in precipitation become more frequent. Our security related infrastructure becomes certainly less adapted.

4 Consequences of Mean Global Warming

Since climate is one of the most important natural resources any climate change is very relevant for all forms of life. Hence, every facet of life will show climate change impacts. Whether for water supply, food production or health it will often have negative consequences. In this chapter, however, only immediate physical consequences of mean global warming will be described. They are mostly related to the water cycle.

4.1 Shrinking of the Cryosphere

The only natural surface which cannot react to a temperature increase caused by an enhanced greenhouse effect of the atmosphere with higher temperatures is a melting snow or ice surface. It will increase the melting rate. Especially during the recent few decades a nearly global shrinking of the cryosphere became obvious also to the laymen. Days with snow cover shrank at most places, permafrost has started to melt in large areas of Alaska and Siberia but also in the European Alps at high altitudes, sea ice cover and extent showed lowest values since observations began in recent years in the Arctic (see also section 2.2), mountain glaciers retreated both by accelerated melting and less snow fall in nearly all mountain ranges, first net mass balance estimates of the Greenland ice sheet using satellite altimetry point to the dominance of melting at the margins over the increased net accumulation in the centre.

The modelling of the cryosphere in climate models is a very difficult task because for sea ice not only thermodynamic processes like freezing and thawing of a salt solution have to be correctly handled but also the drift due to ocean currents and wind forcing as well as deformation by convergent or divergent flow (ice rheology), often leading to packed ice with so-called pressure ridges. Modelling the three-dimensional flow of an ice-sheet is the second major challenge for cryosphere modelling within climate models. The flow field is a function of the geothermal heat flux, trace substance (dust) content, temperature and bedrock orography. First such fully coupled ice sheet models exist, used for assessments of ice sheet geometry in the forthcoming millennia depending on greenhouse gas emissions.

As figure 3.7 already has shown, Arctic sea ice cover will undergo a major retreat in all emission scenarios and multi-year sea ice will disappear in scenarios with CO_2-emissions like A1B or higher. This would eradicate sea-ice ecosystems or at least endanger species depending on the existence of sea ice like Arctic seals and polar bears. Also snow cover for very large parts of Europe and inner Asia will no longer exist into March.

The simulations with an Earth system model including a three-dimensional ice sheet model for Greenland and Antarctica (Vizcaino, 2006) show for the first time that – depending on the scenario chosen – the Greenland ice sheet could melt away nearly completely in the coming few millennia leading to a sea level rise of several meters. This threat for coastal cities and populations can only be avoided if major emission reductions are implemented within the coming few decades.

4.2 Changes in Sea Level

Global mean sea level depends on several factors with very different time scales. It depends on tectonics, i.e. the ocean bottom topography and the percentage of the Earth's surface covered by the ocean varying on time scales of many millions of years, it varies with water stored as ice on continents (≤ 0.35 m sea level equivalent for mountain glaciers, 6 to 7 m for the Greenland ice sheet, about 65 m for the Antarctic ice sheet, with time scales from decades to many millennia), it is a function of ocean water density hence mainly temperature but also salt content. Figure 4.1 is the best available mean sea level rise observation on the basis of satellite altimeters. Compared to the average sea level rise estimate of less than 2 mm/year for the 20th century it shows acceleration to 3.2 mm/year.

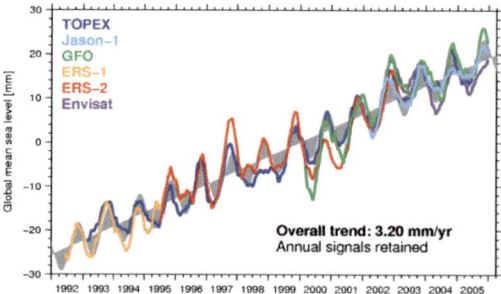

Figure 4.1 **Observed global mean sea level rise since 1992 when continuous satellite altimeter measurements started. The evaluations of different groups using different altimeters are jointly plotted (ESA, 2006)**

In addition, figure 4.2 also shows an annual cycle caused by the warming of the upper layers of the ocean being stronger in the Northern Hemisphere summer. As figure 3.6 already has shown the effect of thermal expansion due to a warming of ocean water is not at all a negligible part of future sea level rise. In 2100, following scenario A1B, the rise is strongly regionalized due to both the different warming of the ocean interior and the re-arrangement of atmospheric circulation in a changed climate. In the European marginal seas, like the North Sea and the Baltic Sea, the rise would be around 0.4 m until 2100 driven by the deep reaching warming during winter storms in the northern North Atlantic and an intensified westerly flow with some additional pile-up of water at the West European coasts.

4.3 Changed Precipitation Distribution

For many regions the amount of precipitation is the single most important climate parameter. Therefore the shift of rain belts as a consequence of unequal warming patterns is the information sought. The calculation of precipitation on the other hand is much more difficult than for temperatures. Nevertheless, some progress was made in validation of climate models with respect to precipitation as a function of season and orography, mainly by improving the spatial resolution of the model both for global and nested regional models.

As already shortly mentioned in section 2.2, the observed precipitation changes are:

1. More precipitation in high northern latitudes throughout the year
2. Reduced amount in rainy seasons for many semi-arid regions both in the tropics and subtropics
3. Tendency for more precipitation in the inner tropics but still not highly significant
4. Higher rainfall amount per precipitation event, a mere consequence of the Clausius-Clapeyron equation at higher surface temperatures, except for areas with strong overall reduction in total precipitation amount.

What will the future hold, if emissions of greenhouse gases continue? In parts it was already displayed in figure 3.5 where the overall effect besides a general increased drying of already dry regions was the general message. What is, however, also an important message for societies, namely the changed probability for floods and droughts, cannot be derived from figure 3.5. Therefore, figure 3.8 had already displayed the changed highest 5 day precipitation amount of a year. Except for North Africa and some other smaller areas with Mediterranean climate situations causing major flooding or very high snow pack in high latitudes or altitudes become more severe. As the climate model used for these projections does only resolve on a 2° x 2° grid, looking for extreme precipitation of a day is not the proper way as such extrema are also depending strongly on orography, which is not properly resolved in such a coarse global model.

If the amount of precipitation per event increases but the total amount of precipitation does not grow to the same percentage level, the average period without precipitation must become longer. Therefore we could expect also an increase in the length of the longest period with precipitation below a certain threshold. This is the case, as figure 4.2 convincingly shows. Large areas, but especially the Mediterranean, may suffer from strongly prolonged drought periods. Figures 3.8 and 4.2 combined are a serious threat for southern Europe.

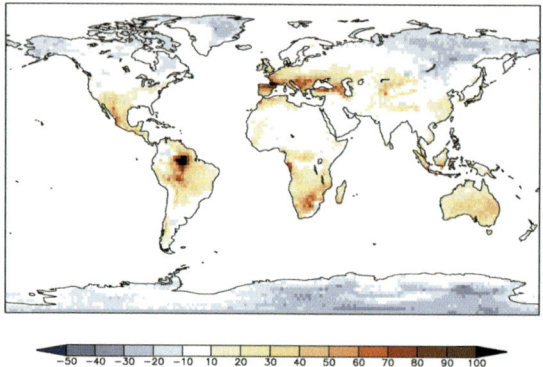

Figure 4.2 Percent changes in the length of the longest period with precipitation below a 10 mm threshold, given in days, for scenario A1B and the period 2071 to 2100 (MPI, 2006b)

4.4 Detection of Climate Change and Attribution of Causes

When the first full assessment report of the Intergovernmental Panel on Climate Change (IPCC) was presented at the Second World Climate Conference (SWCC) in Geneva, Switzerland, in late October 1990, the Ministerial Conference at the end of SWCC in early November 1990 asked for a United Nations Framework Convention on Climate Change (UNFCCC) to be ready for the United Nations Conference on Environment and Development (UNCED) in Rio de Janeiro, Brazil, in June 1992. At this time the urgency for action was not based on the detection of anthropogenic climate change in observations of climate parameters but merely on the following three pillars of argumentation:

1. Rise of long-lived greenhouse gas concentrations: nearly 0.5 percent per year for carbon dioxide (CO_2), about 1.0 percent per year for methane (CH_4) and 0.25 percent per year for nitrous oxide (N_2O) for the 1980s.

Detection of Climate Change and Attribution of Causes   47

2. High correlation between CO_2 and CH_4 concentrations on one side and temperature at precipitation formation on the other, detected in Antarctic ice cores, dating back until about 160,000 years. CO_2 and CH_4 concentrations were derived from air bubbles within the ice core at station Vostok on the West Antarctic Plateau.
3. Projections of climate change in general circulation models of the atmosphere, which were run under conditions of enhanced CO_2 concentration with a shallow ocean mixed layer as lower boundary. The sensitivity to CO_2 concentration increases was judged to lie between 1.5 and 4.5°C mean global warming at the surface for a doubling of pre-industrial CO_2-concentration from 280 to 560 ppmv.

4.4.1   What is detection of anthropogenic climate change?

Detection is the process of demonstrating that an observed change is significantly different (in a statistical sense) than can be explained by natural internal variability (IPCC, 2001a). Hence detection is a signal-in-noise problem. The following ingredients are needed: a) Observations of a climate parameter with high signal-to-noise ratio; b) an estimate of the expected change and c) an estimate of natural variability of the parameter under investigation. The parameter best suited is near surface air temperature T_s, because its measurement was already nearly as precise in the 19th century as today (typically a single reading is accurate to 0.1°C). The expected global annual mean change of T_s could be derived from coupled ocean/atmosphere models for a given greenhouse gas concentration change, e.g. about 30 percent CO_2 increase in 2000 with respect to a concentration of 280 ppmv in 1750. The natural variability on timescales from interannual to a few decades could also be assessed from the observations and the coupled models. With the advent of coupled atmosphere/ocean models in the early 1990s all ingredients were thus available. In addition, the statistical method of searching for the optimal fingerprint (Hasselmann, 1993) was able to enhance the signal-to-noise ratio for a given temperature change pattern knowing the internal natural variability pattern.

4.4.2   What is an attribution of anthropogenic climate change?

There are several pathways of mankind to change climate, e.g. depletion of stratospheric ozone, enhanced turbidity of air by emission of particles and/or their precursor gases, emission of long-lived greenhouse gases, change of surface properties by land use change (construction of roads, changes in crops, deforestation and afforestation, etc.). If the processes leading to climate change are understood and forcing history is known as well, the influence of a distinct human activity can be separated, i.e. a facet of climate change can be attributed to a certain cause. A definition of attribution following IPCC (2001a) is: Attribution of anthropogenic climate change is understood to mean a) detection (as defined above), b) demonstration that the detected change is consistent with a combination of external forcing including anthropogenic changes in the composition of the atmosphere and natural internal variability and c) that it is not consistent with alternative physically plausible explanations of recent climate change that exclude important elements of the given combination of forcings.

Examples for attribution questions are:

1. Do clouds have properties reacted to the recently reduced air pollution in Central and Eastern Europe after the collapse of the East Block and has this accelerated warming at the surface of the region?
2. Is the observed strong cooling trend of the recent decades in the lowest stratosphere caused by ozone depletion rather than by the enhanced greenhouse effect?
3. Does air pollution lead to more or new extreme precipitation events?

4.4.3   First detection of anthropogenic climate change

In March 1995 at a press conference in Hamburg, Germany, at the Max Planck Institute for Meteorology its director Klaus Hasselmann reported the detection of the anthropogenic climate change signal at the 95 percent significance level (Hegerl et al., 1996). The basis of this statement, which soon got support from other groups, was:

Forcing a coupled atmosphere/ocean/land model by reconstructed greenhouse gas concentration time series and comparing the simulated temperature change patterns with observed ones, using the new so-called fingerprint method, similarity of the emerging change signal with the observed pattern pointed to changes going beyond natural climate variability.

Working Group I of IPCC, when meeting in Asheville, North Carolina, USA, in July 1995 for the establishment of a final draft for the Second Assessment Report (SAR) of IPCC and an overall conclusion set the stage for the leading sentence of the SAR:

The balance of evidence suggests a discernible human influence on global climate.

This sentence became the foundation for the Conference of the Parties (COP) to UNFCCC to formulate at COP3 the Kyoto Protocol.

Five years later, end of 2000, the Third Assessment Report of IPCC has strengthened the summary concerning the detection of anthropogenic climate change by stating that new and clearer findings support that most of the warming in the recent 50 years is due to human activities.

4.4.4 Attribution of climate change to causes

The recent years saw a multitude of attempts to attribute observed climate parameter changes to either natural or anthropogenic forcing. One of the earliest attributions was: Cooling of the lower stratosphere (around 20 km height) is due to the depletion of the stratospheric ozone layer by decay products of the chlorofluorocarbons (CFCs) and not so much due to the enhanced greenhouse effect, which also leads to cooling of the stratosphere. By the research group of one of the authors (H. Grassl) several regional attribution attempts have been published recently, showing firstly that the decrease of air pollution over Europe and the increase over China have led to cloud property changes.

While the reflectivity of clouds was reduced over large parts of Europe from the 1980s to the 1990s, because of lower aerosol particle concentrations as a consequence of both emission reduction by new environmental laws in western European countries and collapse of parts of industry in former East Block countries, it was reduced over China also, but because of strong pollution growth with a large soot proportion making clouds reflecting less (Krüger and Grassl, 2002; Krüger and Grassl, 2004). In contrast to it, clouds over major European harbour areas did not show reduced but enhanced reflectivity because the emissions are not regulated and ship traffic was growing fast (Devasthale et al., 2006).

5 Impacts of and Adaptation to Climate Change

As climate is an important if not the most important natural resource any climate change must have an impact on life in all its forms. Because climate must always change or vary due to changes of external forcing and strongly different timescales of the interacting climate system components, all life forms have developed a capability for adaptation to changes, however, only within certain boundaries of climate change. If these boundaries are surmounted either migration or – in extreme cases – extinction is the result. This statement is valid for all life forms. A key argument in the climate change debate is the high change rate going beyond adaptive capacity for many ecosystems and thus species if mankind continues to grow and emit per capita like in recent decades.

Therefore the rapid climate change projected for the 21st century even with major climate protection efforts is an undisputed threat to biodiversity. As stated in IPCC (2007b): There is medium confidence that approximately 20-30% of plant and animal species assessed so far are likely to be at increased risk of extinction if increases in global average temperature exceed 1.5 – 2.5°C over 1980-1999 levels. As shown in section 3.4 global mean anthropogenic warming of the same order of magnitude occurring naturally within many millennia is projected for the 21st century. Hence, major adaptation measures will be necessary in the coming decades. This holds even under future stringent climate change mitigation efforts because of the inertia of the climate system components ocean and parts of the cryosphere (ice sheets, permafrost). Warming of the coming few decades and sea level rise for the coming several decades to a century is already largely pre-determined by the history of greenhouse gas emissions in recent decades. The coming generations have thus to shoulder two tasks at once: Adaptation to and mitigation of climate change.

5.1 Vulnerability

Impacts of climate change on humans depend strongly on the vulnerability of a society or parts of it or a group of persons. If an old village or city behind a dyke in an estuary grows and sea level has risen, the same storm intensity distribution as before makes the settlement more vulnerable to storm surges and endangers more people more strongly. The dyke – maybe constructed to withstand the so-called one hundred year event – must be strengthened to reach the earlier lower risk level again. If, in addition, the storm frequency and intensity distribution changes with climate change a major risk growth will occur, making the population much more vulnerable. If at the same time the economic values in each house grow, there will be a damage explosion as reported recently by insurance companies for all weather-related disasters, in this case when dykes break or are overrun by a storm surge.

It is an empirical fact that the poor parts of a society and the poor countries are much more vulnerable to natural and weather related disasters than the more developed countries or the wealthier parts of the society. Climate change is thus – via enhanced vulnerability – a key equity problem of global proportions. This problematique has implicitly been accepted in the UNFCCC and its Kyoto Protocol by asking the industrialized countries for emission reduction commitments first.

5.2 What is a Climate Change Impact?

The diversity of impacts of climate change in different sectors of society does make it difficult to give a simple definition. Therefore, it is best to describe firstly the different impacts per sector and – where necessary – secondly the combination of impacts on a certain sector. An example for the first is the higher probability of damage by storm surges just because of local sea level rise to which the global mean rise has contributed. The second example is the combination of sea level rise with enhanced storm frequency and/or intensity causing a further increase of vulnerability without adaptation measures. A further example for the second case is a redistribution of precipitation within the year towards winter coming on top and leading to an ever higher storm surge level in the upper part of an estuary.

If the different impacts within a country are summed up, the aggregation of sectors will lead to a relative gross national product (GNP) change, which, however, would no longer detail the often strongly negative impacts because also positive impacts are now included.

Examples are: damage to agriculture in more frequent exceptionally hot and dry summers versus increased income by enhanced tourism; less heating oil consumption in generally milder winters with more frequent and intense storms; die-back of certain tree species because of more probably summer drought versus higher quality of wines in the northern parts of wine-growing climates.

What is a Climate Change Impact?   53

5.2.1   Impacts on sectors

Climate is a basic natural resource. Hence its change will have an impact on all natural and managed ecosystems as well as on all parts of human societies. Therefore the UNFCCC main goal speaks of the avoidance of a dangerous anthropogenic interference with the climate system and specifies what sectors should keep adaptability to climate change: natural ecosystems, agriculture (food security), economy (proceeding in a sustainable manner).

Natural ecosystems

The key point here is the ability of forests and other highly vulnerable ecosystems to adapt to climate change. Therefore, the climate change rate is the decisive parameter. However, there is no sufficiently detailed research available to give a threshold climate change rate. IPCC (2001b) has given instead estimates of maximum warming in the 21st century for different types of natural ecosystems (see figure 5.1), showing that a mean global warming going beyond 2°C (in comparison to the pre-industrial value) is very likely a dangerous interference with the climate system. Climate change going beyond the threshold would lead to a threat to biodiversity that is probably as large as the one caused by direct habitat destruction.

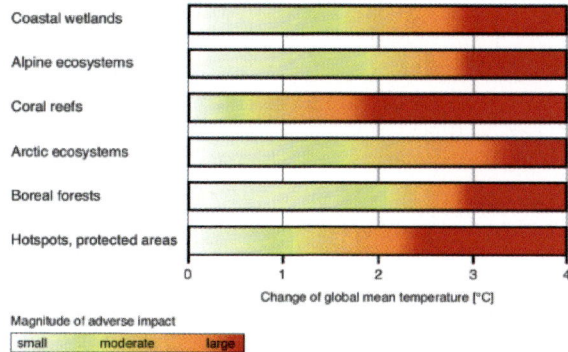

Figure 5.1   Vulnerability of different ecosystems to a global mean temperature rise, given in estimated damage levels

Forestry

Foresters in mid-latitudes and higher latitudes deal with organisms that can easily reach an age of 100 years. Therefore, the projected climate change in the 21st century – even with strong climate protection measures – poses unprecedented forest management questions: Which trees should not be planted because they would be attacked by pests and diseases too early to be of value? Should one introduce new tree species from warmer climates anticipating the projected warming and the probably drier summers? Or should trees from the southern part of its natural habitat be preferred?

Managed forests are the managed ecosystems most strongly impacted by climate change. Foresters in Central Europe observe already now the climate change impact when key tree species are no longer able to withstand known pests and diseases or invading ones. Another major point for the climate change debate is expressed in the following question: At which warming and precipitation change will parts of the carbon stored in forest and their soils be emitted into the atmosphere? At present a considerable portion of anthropogenic CO_2 is stored additionally in northern hemisphere forests because of the CO_2 fertilization effect.

Agriculture

Modern agriculture in many OECD countries is producing more food than needed in these countries and its management is depending less on climate change than on agricultural subsidies that amount to about 250 billion Euros per year worldwide and which hamper the export of agricultural goods by developing countries. Subsistence farming, on the other hand, is the most vulnerable agriculture to climate change. Several hundred million people thus are suffering and will increasingly suffer from climate change, caused mainly by industrialized countries and with growing proportions by emerging countries. This inequity is a major reason for tension during the conferences of the parties (COPs) to the UNFCCC. While farmers in developed countries can easily cope with changed crops and are less hit by climate change (IPCC, 2001b) the poorest farmers in developing countries will have to give up more and more soils if semi-arid areas become even drier, as climate models suggest.

Marine ecosystems

These ecosystems depend on ocean currents, pH-value of ocean water, nutrients, insolation at the surface as a function of season, and in parts on sea ice cover and type. As all these parameters change systematically and also their frequency distribution might change, major changes can be expected during the 21st century. The recent rapid warming, for example in the North Sea, has already led to a shift in biodiversity with increasing numbers of species from warmer ocean climates invading and others leaving to higher latitudes. Because overfishing is common in nearly all ocean basins, and since long-term systematic observations are nearly non-existing, an assessment of climate change impacts is often very difficult. As an example of the change of a fundamental chemical parameter, figure 5.2 shows present pH-values measured, their values during the last glacial and those anticipated at a certain CO_2-concentration in the 21st century. Please note this change is a mere consequence of changed atmospheric CO_2 concentration.

Water supply

Water is a basic necessity for life. Its redistribution with global climate change represents the single most important consequence of climate change besides

What is a Climate Change Impact? 55

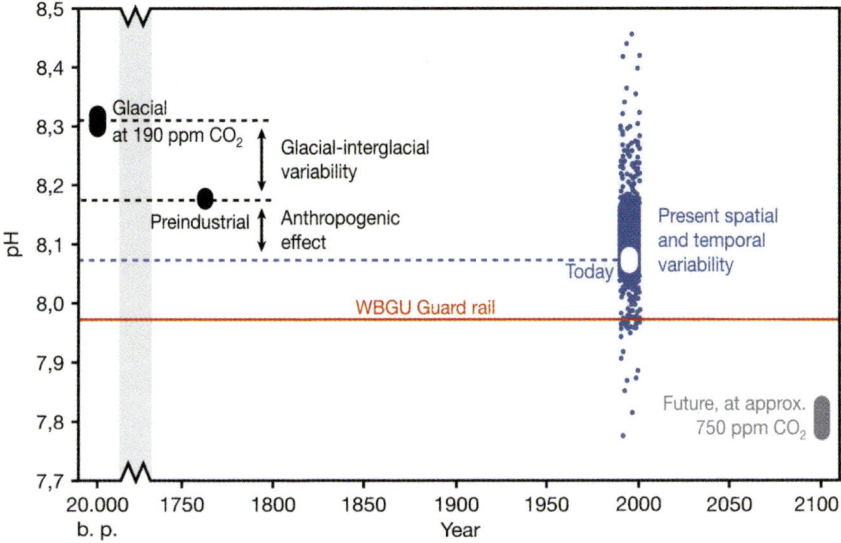

Figure 5.2 Measured, reconstructed and projected pH-values of sea water together with a threshold (guard rail) value not to be trespassed in order to avoid major damage to marine ecosystems (WBGU, 2006)

sea level rise. As always also here local mismanagement (polluting water and not treating it afterwards) combines with a global change facet and leads to an increase of water scarcity in many, especially semi-arid countries. As an indication of intensified drought conditions figure 4.2 has already displayed the increase in days of the length of dry spells in scenario A1B at the end of the 21st century. It is obvious that especially Mediterranean climates and inner parts of continents outside high latitudes will experience more severe water shortages.

Security infrastructure

Mankind has always adapted to climate variability, hence to weather extremes, depending on its economic strength. Therefore, developed countries have better adapted construction codes, coastal defence, and flood protection. However, this once adapted infrastructure becomes ill-adapted, if frequency distributions of climate parameters change both mean value and shape. Hence, all countries are increasingly loosing parts of their adaptation, as became obvious during recent weather extremes.

To adapt this security related infrastructure anew is probably the most expensive part of adaptation to climate change in the coming decades. On the other side, much less expensive is mitigation of climate change by reduction of emissions, if started now (Stern, 2006), because adaptation costs to climate change would rise to levels beyond 5 percent of GNP, if no action to reduce the climate change rate is taken.

Human health

The climate parameter temperature is key for all vector-borne infectious diseases that still are one of the main causes for death, especially in developing countries. Countries where malaria is common, e.g. lowlands of Kenya, will see an extension of malaria into elevated areas; other tropical or subtropical infectious diseases will creep northward and southward with global warming.

Further threats to human health are heat waves. The heat wave in Europe in summer 2003 has caused at least 35,000 deaths not only among elderly people but also in other age groups. However, it would be easily possible to reduce this burden by better information and adaptive measures. Heat waves uncover the "social freeze" in societies. Cold spells will certainly become less threatening for human health.

An area with potential health impacts is the changed composition of food grown under changed climate conditions. However, not enough is known yet to assess its severity.

Economy

While the last decade was filled with the claim by scientists that potential climate change mitigation measures are much more costly than adaptation, the situation has changed drastically. As expressed for example by the German Global Change Advisory Council in its report to the government entitled "Towards a Sustainable Energy System" (WBGU, 2003a), a strongly growing global economy following the multilateral approach and being fond of innovative technology (i.e. scenario A1T with stabilization at 450 ppmv CO_2 concentration) would not cause a burden going beyond 1.5 percent GDP for any region on the globe. However, as recently published by Stern (2006), unabated climate change could drive global economy into adaptation measures going even beyond 5 percent of global economic turn over.

5.2.2 Impact on certain geographical regions

It is clear that a hyper-arid region remaining hyper-arid at higher temperatures will not experience a major climate change impact. If, however, a mountain range in the midlatitudes looses its glaciers a major water supply shortage in summer and autumn is the immediate consequence. One of the most important climate change impacts will be felt at nearly all coasts, namely coastal erosion, due to sea level rise, which will also cause more frequent inundation of marshlands even at unchanged storm frequency and severity.

Marshlands and coasts

About 50 percent of the world population lives close to coasts (within 50 km) and many multi-million cities stretch along coasts or estuaries and within marshland areas. For the latter sea level rise often threatens their mere existence. While the Netherlands have a nation-wide policy to protect the entire coastline by sluices, dunes and dykes at least 15 m above mean sea level, German coastal states have agreed to protect themselves against a 100 year storm surge by infrastructure reaching about 8 m above mean sea level. Most cities in developing countries or small island states in the tropics are far from such a protection level. Climate change-driven sea level rise, therefore, is the best example for "ecological aggression", the term used by the former executive director of UNEP, Klaus Töpfer, with respect to climate change caused by the rich North. A large part of adaptation to climate change will circle around coastal protection.

Permafrost areas

Many countries also outside high latitudes, e.g. China, have permafrost underlying a rather large percentage of their territory. Therefore, melting of permafrost is a problem for most countries in the world, including subtropical and even tropical countries with high mountains as well.

The first reaction to a systematic warming in permafrost areas will be a deeper active layer without frost during the late summer half year. This will cause a growing collapse of infrastructure like roads and buildings because deeper melting will reduce the unevenly distributed ice lenses unevenly. Subsidence of the soil is thus irregular, destroying many buildings and calling for frequent reconstruction of roads.

Thawing permafrost in mountain ranges is often the nucleus for major landslides during summer-time heavy precipitation events, the movement of avalanche protection and the outburst of glacial lakes.

Mountain areas

Mountain ranges are not only the water towers for the low-lands but also biodiversity hot spots and regions with especially frequent weather related disasters, like avalanches, land-slides, flash floods, wild fires. Warming alone, without a change in the precipitation regime, will change the water cycle in mountain areas strongly because of the upward shift of snowfall by about 150 m per °C. Therefore, the annual hydrograph of rivers will change, water for irrigation as well, small glaciers will melt completely. If the precipitation regime will also change, certainly the standard case, agriculture in mountain areas and adjacent lowlands may change dramatically. Climate models for the 21st century scenarios indicate a drying of the subtropical mountain ranges from Northern Africa and the Mediterranean area into Central Asia as one of the key results, translating into severe water shortage at no or only small reduction of flash flood severity as the surface temperature rises. How people in the mountain areas and adjacent lowlands will react to the water shortage is probably the following: drilling deeper and using more fossil water or – for oil countries – desalination of sea water by using more oil. This is, however, unsustainable. Major strategy changes in water management are needed.

6 Sustainable Development and Climate Change

Since 1987, when the report of the World Commission on Environment and Development entitled "Our Common Future" was submitted to the United Nations, in response to a request by the UN, the term "sustainable development" has been coined for the long-term development strategy of mankind. Its comprehensive and simple definition is: Sustainable Development is development that meets the needs of the present without compromising the ability of future generations to meet their own needs. Despite the many re-definitions, also by interest groups, this term remained the leading one for the United Nations, which tried to foster such a development through the Millennium Development Goals in 2000 and the action plan of the World Conference on Sustainable Development (WSSD) in Johannesburg, South Africa, in September 2002. It is easy to define lack of sustainability, for example sealing of soils in Germany, where the population has started to decline, but it is difficult to define the corridor to reach "Sustainable Development". An example for this difficulty is: At which climate change rate will already ongoing biodiversity loss be accelerated so much that the rapid transition to a largely different and threatening climate state will be triggered by the emergent instability of ecosystems? The answer cannot be given yet!

6.1 The Enhanced Greenhouse Effect without Analogues in Climate History

Going beyond chapter 2, we introduce here the uniqueness of the present state of the Earth system from the point of view of its physical boundaries. From greenhouse gas concentrations derived from air bubbles in ice cores drilled into the ice sheets of Antarctica, but also Greenland, down to the bedrock we know that the second most important greenhouse gas, CO_2, was varying between about 190 ppmv during glaciation maxima and slightly less than 300 ppmv during the interglacials, often called warm periods. Hence, the present CO_2 concentration, already above 380 ppmv, never occurred since at least 750,000 years. We are therefore in a non-analogue state for homo sapiens. Can we estimate the future climate then although climate history gives no analogues? In principle yes, but only with Earth system models validated with climate history, thus comparing them with so-called proxy-data. Only very recently did such testing become possible, but mainly for Earth models of intermediate complexity (EMICs), because the computer time requests of higher spatial resolution models are still too high. As EMICs and first fully coupled higher resolution models with 3-D ice sheet representation have shown that bifurcations are close for business as usual scenarios of human behaviour in the 21st century, sustainability is threatened by further climate change or in other words a stringent global climate policy is needed to approach sustainability in the 21st century. As figure 6.1, taken from Vizcaino (2006), demonstrates, sea level rise up to several meters would follow the continuation of the present development path, where energy supply is based on fossil fuels. Many millions, living in marshlands at unprotected coastlines, would then become refugees. Strange as it may seem, a stop of the North Atlantic Drift (vulgo Gulf Stream) would, it occurs for scenario A2 and for some realisations already for scenario A1B, reduce sea level rise, because more ice on Greenland would continue to exist. It is clear from figure 6.1 that our climate policy in the next decades decides about sea level rise in centuries and millennia to come and thus about the further existence of coastal cities without very costly major local protection measures.

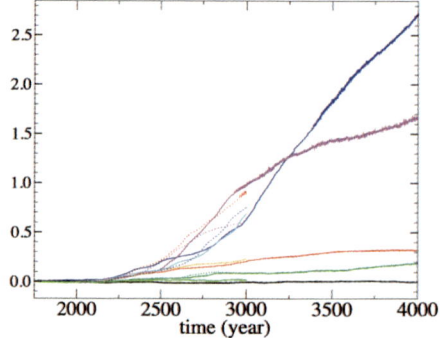

Figure 6.1 Melting of the Greenland ice sheet, given in meters of sea level equivalent for several realisations of scenarios A2 and A1B during the coming millennia. Please note that emissions are decaying after the year 2100 or 2200 (for A2) exponentially but global warming and sea level rise continue over centuries and millennia (Vizcaino, 2006)

6.2 Carbon Cycle Feedbacks

The reservoirs soil (~1200 GtC) and vegetation (~500 GtC) together contain more carbon than the atmosphere (~760 GtC). Therefore, climate change can lead to considerable feedbacks of these two reservoirs. As already stated in section 2.5, the terrestrial biosphere stores ~2.5 GtC per year in addition at present, i.e. helps to reduce CO_2 growth rate in the atmosphere. Model results concerning the strength of this sink or even its change into a source are inconclusive. A further potentially strong feedback is related to the melting of permafrost as this frees soil layers containing a lot of carbon that can be converted – depending on local conditions – into CO_2 and/or CH_4. Our knowledge about the melt rate is insufficient to call it a reliable part of an Earth system model.

Often the huge carbon reservoir in the ocean (~39,000 GtC) is seen by laymen as a major potential CO_2-source if the ocean warms. However, the carbon chemistry in the ocean is comparably well understood to rule out a major positive feedback due to outgassing at higher sea (surface) temperatures. Reactions by phytoplankton, both related to total biomass production and changed species composition, pose more difficult questions. The reaction by phytoplankton also depends on the reduced pH-value of ocean water (see also figure 5.2) because some phytoplankton species forming carbonate shells are affected by lower pH-values.

6.3 Sequestration of Carbon

Carbon containing fuels (peat and so-called fossil fuels) were formed by sequestration of carbon containing detritus of organisms into bogs and sediments. Coal, oil and natural gas have been formed over many millions of years from this detritus and were stored often deep in the Earth's crust. Therefore, pressing the carbon dioxide into former oil and gas containing layers as well as coal seams has been proposed as a climate change mitigation measure. It has already been used in the USA (Texas) to enhance oil recovery. Estimates of the storage capacity (see Haszeldine, 2006) range into hundreds of GtC, in form of liquid CO_2 that, deeper than 800 m below surface, fills as a hypercritical liquid the pores of the layers that contained oil and gas and contain at present saline water, and mixes with it. The key question will be: Is the additional energy needed to separate CO_2 from the flue gas, the transport to the "burial" site and pressing it into the former oil and gas deposits small enough to stay economic vis-à-vis renewable energies with steep learning curves in an environment of political support? For more details see chapter 9.3.

In the context of sequestration two other strongly differing approaches are also discussed: Firstly, sequestration of CO_2 resulting from power plants using recent biomass, either in the form of remnants from agriculture, or energy plantations (rape seed, sugar cane, miscanthus, rapidly growing trees), and secondly creation of soils in the tropics containing large amounts of charcoal (terra petra in Amazonia), produced in biomass power plant installations as well (Lehmann et al., 2006).

6.4 Barriers, Opportunities and Market Potential for New Technologies and Practices

Always, when a major restructuring of economy as well as behavioural changes are needed, many potential and some real loosers will oppose. As the "old-fashioned" industries are typically well organized, new economic opportunities, which often need government support for the introduction into the market, will not get it. The "old" will be able to persuade not to change the incentive structure for some years, until the new industries will have shown their superiority at least in niche markets.

We will not deal with all the barriers for climate policy as a major hindrance for our approach to sustainable development here in detail but a very short history of arguments against climate policy measures will nevertheless be given in table 6.1 in the form of a comparison of two types of statements, one by mainstream science and the other by interest groups in order to show the "learning" curve of interest groups.

Table 6.1   Comparison of statements as an example of one of the main barriers to reach an effective climate mitigation policy

Time	Science argument	Typical arguments of interest groups
Late 1980s	There is a mean global warming since the beginning of the 20th century whose origin is still not clear; climate model estimates point to several degrees warming in the 21st century if climate policy measures are not taken	Measurements with high error bars including urban heat islands cannot be trusted; climate models are unable to project global warming correctly
Mid 1990s	The anthropogenic warming signal has been detected in observations	The warming (now admitted) is still part of natural variability
2000	The major part of the mean global warming during recent decades is anthropogenic, while the warming until about 1940 is in parts also caused by higher solar irradiance	A warmer planet is a greener one and adaptation to climate change is less costly than mitigation policies
2006	Adaptation costs incurring without climate protection measures strongly surmount mitigation costs, hence action now is an insurance policy	The economic models used for these cost estimates are not credible

But also in the area of opportunities existing for an overhaul of the fossil fuel based energy supply system of entire mankind often chances are exaggerated.

Therefore, the opportunities section has to start with a description of the technical potential of different renewable energy resources.

6.4.1 Basis of opportunities: The technical potential of renewable energy

The proper parameter to describe the technical potential of renewable energy forms is the energy flux density, here given in Watts per square meter (Wm^{-2}). A densely populated and developed country like Germany has an energy supply system with about 1.5 Wm^{-2} primary energy flux density, of which about 90 percent are fossil fuel based. In comparison, the geothermal energy flux density reaches only 0.08 Wm^{-2} on average in Germany. Hence, geothermal energy can only become a small part of a renewable energy supply system. In addition, the comparably low temperatures of only several hundred °C make geothermal energy not very attractive for electric energy generation (only about 20 percent efficiency). Therefore, geothermal energy will mainly be useful for heating of buildings and industrial processes running at higher temperatures.

For an overview of renewable energy forms a ranking of technical potentials from the smallest to the strongest contribution is a first useful assessment. From table 6.2 it is becoming evident that in the long run the use of direct solar radiation is the only nearly inexhaustible energy source, if one compares it to the present and future energy demand by humankind: less than 0.03 Wm^{-2} today and about 0.08 Wm^{-2} in 2100. The latter number assumes development of developing countries in the 21st century and a continuous increase in energy productivity of 1.5 percent per year, up from 1.0 percent/a observed in the 20th century. A major hurdle for the introduction of some renewable energy sources as a component of the energy supply system is their intermittency. Therefore, research for storage of intermittent energy forms e.g. solar energy, is a major point besides development of large electrical grids, if they do not yet exist.

Table 6.2 Ranking (from smallest upwards) of the technical potential of renewable energy resources given in global average energy flux densities (Wm^{-2})

Name	Energy source	Technical potential	Availability
Tidal energy	Gravitational force of Moon and Sun moving masses	< 0.01	Intermittent but regular
Water wave energy	Atmospheric wind energy	< 0.01	Intermittent

Name	Energy source	Technical potential	Availability
Hydropower	Potential energy of water in higher elevation	< 0.01	Intermittent to continuous
Geothermal energy	Upward heat flux in the Earth's crust	~ 0.1	Continuous
Biomass energy	Formation of carbohydrates by plants from solar energy	~ 0.1	Nearly continuous
Wind energy	Near surface atmospheric pressure gradients caused by unequal solar irradiance	<< 3 *	Intermittent
Solar irradiance	Nuclear fusion in the sun	~ 165	Intermittent

* ~ 3 Wm^{-2} is the entire kinetic energy of the atmosphere.

6.4.2 Development status of renewable energy sources

The support of new energy technologies is strongly depending on the availability of a certain renewable energy resource in certain areas or countries. Examples are: Norwegians have to heat houses with electricity from hydropowerplants. The Icelandic heating of houses uses geothermal energy. Warm water is generated with solar panels on roofs in Israel. District heating comes from wood-based heating plants in some forest-rich parts of Austria. Solar thermal power plants in deserts produce electric current needed during peak air conditioning demand during noon and afternoon in California. As all the above examples and many more show, renewable energy markets are still often niche markets and their development needs strong incentives from the politicians, because most energy markets are strongly distorted by subsidies for fossil fuels. A good example is the prize of natural gas exported by the Russian Federation to OECD countries which is by more than a factor of 4 above the prize in Russia.

The strongest incentive for renewable energy markets would be the end of direct subsidies for fossil fuels and the internalisation of external costs. Since these measures would reduce the power of some major players, resistance for an ecological tax reform is very strong and only first steps have been taken in some countries. Sometimes it is easier to directly support the development of renewable energy markets, when they are still very small leading to steep cost reductions because of steep learning curves for new technologies and thus reduced necessary support. As with all new technologies, government support for research is needed also for market introduction. Again mighty players handling older technologies will oppose.

Of all modern renewable technologies wind energy is closest to profitable prices without direct support and would often be already profitable, if correct prices for fossil fuel based power plants without externalities covered by tax payers would exist. Also solar thermal power plants in lower latitudes belong to this category.

A major challenge is storage of renewable energy. From table 6.2 it is evident that the strongest renewable energy sources are intermittent. While photovoltaic cells react nearly immediately to cloudiness, solar thermal power plants react in a delayed mode, updraft power plants would deliver into hours after sunset. However, a mix of several renewable energy sources and electric current prizes depending on generation level would in parts offset the need for storage capacity.

Renewable energy for the mobility sector, for example oil from seeds or gaseous and liquid hydrocarbons from biomass, is still a niche market in most countries, except Brazil. It does contribute to the CO_2 emission reduction goal but continues with the air pollution problem when burning hydrocarbons and does not stop nitrous oxide (N_2O) emissions. Only a hydrogen-based energy system for traffic would avoid the air pollution problem as well but exclusively if hydrogen would be produced from renewable energies.

Details on climate effects of biomass derived fuels are discussed in chapter 13.5.

6.4.3 Emissions Trading as a New Practice

The Kyoto Protocol (see also section 7.3) has introduced new international policy instruments.

1. Emissions Trading
2. Joint Implementation
3. Clean Development Mechanism
4. Accounting for Carbon Sinks

Here emissions trading will be described with a bit more detail as it could become the main driver of climate change mitigation policies. It is a market instrument, allocating the full or large parts of the CO_2 costs paid by countries not reaching their targets to the countries surmounting an emission reduction goal. Defining by an international legal instrument – like the Kyoto Protocol – an allowable emission, countries not reducing emissions as prescribed by the legal instrument, can buy (trade) emission certificates from countries having reduced more than prescribed. In the present international political context several former socialist countries, which have achieved strong emission reductions since 1990, can offer these reductions, calculated in equivalent CO_2 emission tons, to some OECD countries that have been unable to reach their emission reduction targets. Hence, in 2012, at the end of the Kyoto period, reductions gained because of a mere political change without any consideration of climate protection, can be traded. This has been called by environmental groups trading with hot air. From the point of view of CO_2 levels in the atmosphere, any reduction caused by whatever political measure counts.

6.5 Costs of Adaptation

Climate change is ongoing and will be accelerated in the near future despite all measures taken by climate policy, as the reaction of the climate system is delayed by decades and centuries. Therefore, adaptation to climate change is a must and will be costly and is in addition to climate change mitigation measures. It is very difficult to estimate adaptation costs. Hence, it is even more difficult to find the minimum costs for a mix of mitigation and adaptation measures. At the same time such a policy goal to find the minimum would be based solely on economics of mostly the developed countries. It would disregard the key ethical problem of global climate change: Suffer will mainly those not having caused it to a large extent.

The recent report by Stern (2006) has facilitated the debate. Adaptation costs to unabated climate change are so high that they would cause a major economic crisis within the 21st century and mitigation costs are much less. Massive mitigation efforts are therefore needed. Although the Stern report did not report new research results but assessed existing ones carefully with a focus on economical consequences, its impact was very high, especially since a high rank economist said it. It is now highly probable that mitigation measures bring a strong economic dividend and – in view of the North-South conflict – less conflict potential.

6.6 Mitigation of Climate Change as the Prerequisite for Sustainable Development

As described in chapter 3, scenarios whose main ingredient is business as usual will lead to a mean global warming going well beyond the experience of homo sapiens and also going beyond adaptability of natural and probably managed ecosystems. Therefore, sustainable development cannot be reached without mitigation of climate change. It is easy to request a major reduction of greenhouse gas emissions, but it is very difficult to negotiate international treaties that lead to emission reductions, because not only companies shy away from it, but also countries whose interests are dominated by fossil fuel production and export or which feel pressure by major oil and coal companies, as clearly dem-

Mitigation of Climate Change as the Prerequisite for Sustainable Development   67

onstrated by the behaviour of USA, Australia (until very recently) and Saudi Arabia in the international negotiations (see also chapter 7).

Without a long-term goal how to reach the corridor of a sustainable development climate change policy remains a piecemeal endeavour. We need clear guardrails.

6.6.1   Guardrails

Guardrails guide our climate change policy. They have to originate from boundaries given by the biosphere and the socio-economic sphere. As figure 6.2 explains, political measures taken can either keep us within the sustainability corridor or bring us back. Examples for one or two measures in each sector in figure 6.2 are:

Example 1 (green sector): Change of heating system using fuel-wood in a forested area into a biomass power plant with district heating using wood and agricultural residues and feeding in the electric current into the public grid. This measure strongly reduces air pollution, reduces fossil fuel use for electric power generation, brings the energy from residues into the energy supply system and generates new income.

Example 2 (red sector): Avoidance of increasing the soils sealed when building new infrastructure by giving back the same area used for the new socio-economic infrastructure from no longer used industrial sites to a natural succession of plant and animal communities, thereby enhancing biological diversity, thus making ecosystem response to climate change more robust.

Example 3 (red sector): Obligatory energetic use of methane emanating from coal mines, which would for example nearly eliminate one of the main meth-

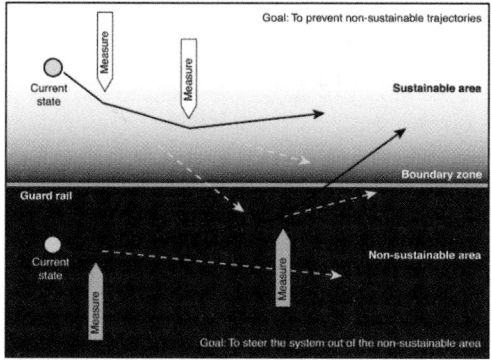

Figure 6.2:   **How to reach a sustainable development path? (WBGU, 2003b)**

ane sources in Germany where coal mine methane is often not burnt when not used (Details see chapter 13.2).

Guardrails have to guide the climate change policy measures. Besides the biosphere driven tolerable climate window (figure 6.3) many other guardrails derived from the sustainable development goal have to be taken into account, for example enough electric energy per person per year (500 kWh) for a decent chance to develop, i.e to reach higher education, get better health care, earn more income. For details on how many guardrails have to be taken into account please refer to the internet address www.wbgu.de from where WBGU (2003a, b) and all other reports since 1993 can be downloaded.

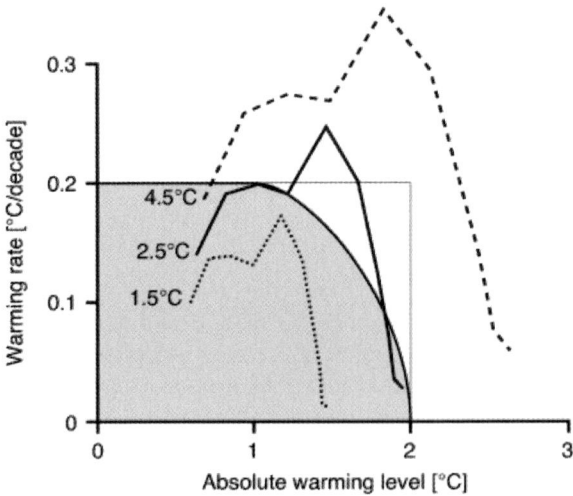

Figures 6.3   The tolerable climate window

6.6.2   Geo-engineering?

If renewable energy is still more expensive than subsidized or non-subsidized fossil fuels, even under an international emissions trading scheme like the one within the Kyoto Protocol (see section 7.3), the difference in prize may stimulate the further use of fossil fuels, however, only if the emissions of greenhouse gases are avoided at least in parts by sequestering carbon dioxide through its deposition underground, preferably in used oil and gas reservoirs. There could also be an incentive to manipulate the atmosphere, if the ecological consequences of the "active climate management" would be known and were less expensive than reducing the main cause of global climate change, the higher concentrations of long-lived greenhouse gases. Both types of measures are known as geo-engineering.

While the first type has already been mentioned in section 6.2 and some pilot projects are under way (see chapter 9.3), the latter would be a first active change of atmospheric composition in order to off-set consequences of the enhanced greenhouse effect. Recently Crutzen (2006) proposed a major research activity to study the consequences of an artificial sulphuric acid aerosol layer in the stratosphere mimicking the situation after a major explosive volcanic eruption, when for some years the Earth surface is on average cooler by several tenths of a degree centigrade. The probability of agreeing after strongly increased knowledge to such a manipulation of the atmosphere, which would need to be a sustained effort over many decades and part of international law, is extremely low.

6.7 A Sustainable Energy Path

In 2003 the German Global Change Advisory Council, asked to give advice to the German government, published a potential energy path, which implements many more guardrails than the one described in section 6.6.1 (WBGU, 2003a). As demonstrated in figure 6.4 it would lead to a sustainable energy system, because mankind could change from fossil fuels as the main pillars of the energy supply to renewables, mainly direct solar radiation, still always taking the cheapest energy source counted over the life time of an installation, and **not** building on behavioural changes of mankind. Sequestration of carbon in former oil and gas deposits is needed for some decades until about 2050 (see chapter 9). This sequestration could be avoided if a more stringent efficiency increase and a more rapid phase out of subsidies for fossil fuels in major countries were implemented (WBGU, 2003a). A detailed description of the current situation and future trends are given in chapter 13.

As discussed earlier in section 6.5, adaptation costs without climate change mitigation measures will run into the above 5 percent of the global economic budget range and mitigation costs are much lower on average. Again in 2003, the German Global Change Advisory Council (WBGU, 2003b) has shown for a subdivision into 11 regions that none would have to invest more than 1.5 percent of GDP in a contraction and convergence scenario aiming at equal emissions per capita in all regions to be reached in 2100, keeping all the guardrails, some of which were described above in section 6.6.1. Figure 6.5 demonstrates that A1T remains the most attractive of all scenarios, also as far as the smallest range in GDP contributions to climate change policy is concerned.

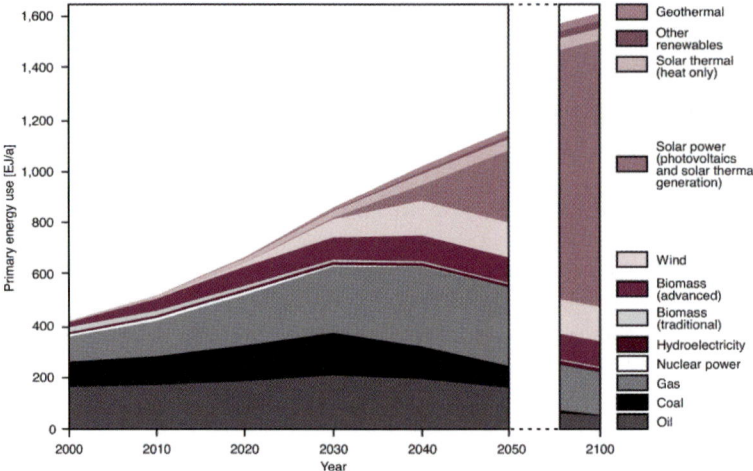

Figure 6.4   A sustainable energy path in the 21st century based on the scenario A1T (see section 3.3)

7 International Climate Policy Approaches

There is no other more global issue than rapid global climate change. Organisms in the deep sea, polar bears and many human societies are threatened by it. Obviously, there must be a global approach to climate policy. Although the knowledge about the greenhouse effect of the atmosphere dates back into the 19th century and the potential concentration change of a major greenhouse gas, namely CO_2, was discussed in the late 19th century, it was only in the second half of the 20th century that observations of a CO_2 concentration change emerged. Several years after the installation of measuring stations on the slopes of a volcano (Mauna Loa) on Hawaii and the South Pole, both the global upward trend and the annual cycle of CO_2 in the Northern Hemisphere became clear. As the period from the early 1950 to the 1970s was one without a mean global warming, mainly due to a strong rise in air turbidity related to sulphur dioxide emissions largely from coal burning – as fossil fuel use grew by more than 4 percent per year globally – the knowledge about the growing CO_2 level did not reach many political decision makers. When did the global political community first react to scientific findings?

7.1 First Policies, Measures and Instruments

The World Meteorological Organization (WMO), a specialized technical agency of the United Nations, called all nations to the First World Climate Conference in 1979 to Geneva, Switzerland. At this conference a World Climate Programme has been initiated, whose research part, the World Climate Research Programme (WCRP) got the task to further the understanding of the climate system, in order to answer the question whether there is a human influence on global climate. The International Council of Scientific Unions (ICSU), now the International Council for Science (with the same acronym), which is a non-governmental organisation, joined WMO in 1980 to launch the WCRP.

7.1.1 The Villach Conferences

In the early 1980s WMO and the United Nations Environment Programme (UNEP) organized the Conferences in Villach in Austria. The third in 1985 issued for the first time a warning about future global climate change if the use of fossil fuels would continue unabated. Their conclusion culminated in a statement that security related infrastructure would then no longer be adapted to climate variability. In other words: known weather extremes would become more frequent and new ones would emerge, as observed now for some parameters like rain amount per event in many areas.

7.1.2 Intergovernmental Panel on Climate Change (IPCC)

As became obvious with the advent of general circulation models calculating global warming near the surface due to an enhanced greenhouse effect in the 1980s in so-called equilibrium runs for doubled CO_2 concentration, an assessment of knowledge on climate change as input information for political decision making was urgently needed. WMO und UNEP therefore formed an intergovernmental body that should be able to give such an authoritative view. The Intergovernmental Panel on Climate Change (IPCC) became right from the beginning in November 1988 such an authoritative voice. Because it was fortunately dominated by leading scientists, who were publishing already in October 1990 the first full assessment report and a summary for policy makers, that set the stage for the later Framework Convention on Climate Change (FCCC). Since then IPCC became through its further regular assessments in 1995, 2001 and 2007 a shining example for proper assessments of knowledge in a science area. As IPCC never intended to give political advice but mere policy-relevant information it was difficult to entrain it into a political debate. IPCC's assessments are used by the Subsidiary Body on Scientific and Technological Advice (SBSTA) of the United Framework Convention on Climate Change (UNFCCC) for political advice to the negotiators (see section 7.2).

Although geographical balance of authors and contributors is sought by IPCC and governments can propose authors and reviewers of chapters IPCC has still managed to attract leading scientists for all working groups. This is fundamental for an in-depth assessment. One of us (H. Grassl) has also been involved deeply until 1995 in the two first assessments and has learned to loose preoccupations during the intense scientific debate in IPCC groups.

7.2 United Nations Framework Convention on Climate Change

Fortunately, the Second World Climate Conference in 1990 requested in its ministerial part a United Nations Convention on Climate Change to be ready for the Earth Summit (United Nations Conference on Environment and Development, UNCED) in Rio de Janeiro. Its official title at signature by 153 heads of States in June 1992 became: United Nations Framework Convention on Climate Change (UNFCCC), indicating that other legal instruments would be needed.

UNFCCC has a central goal in its paragraph 2: *The ultimate objective of this Convention and any related legal instruments that the Conference of the Parties may adopt is to achieve, in accordance with the relevant provisions of the Convention, stabilization of greenhouse gas concentrations in the atmosphere at a level that would prevent dangerous interference with the climate system. Such a level should be achieved within a time-frame sufficient to allow ecosystems to adapt naturally to climate change, to ensure that food production is not threatened and to enable economic development to proceed in a sustainable manner.*

Its implementation is the task of a century, because the time-scales of climate system components reach centuries and even millennia and any clearly visible impact of measures taken now on climate variables will only be seen after several decades. The three side-conditions under which stabilization of greenhouse gas concentrations in the atmosphere has to be reached are not yet fully understood scientifically. Hence UNFCCC is a climate research promoter of highest calibre.

The translation of the above conditions into the challenges ahead can be made clear by the following questions:

1. At what temperature increase rate can forests no longer adapt? In other words: Can boreal forests move within a century by several hundred kilometers northward? West European forests could move in about 10,000 years from southern France to Northern Norway. Under Scenario A2 the same move would be squeezed into about a century. Certainly A2 would lead to the collapse of forest ecosystems, at least to such an extent that their services to us would be drastically diminished.

2. How strong must the precipitation belt shift be to generally endanger food supply for mankind? What can be tolerated? More food in mid and high latitudes while a serious food crisis leads to migration out of enlarged semi-arid zones, already suffering from desertification now?

3. At which total costs (adaptation plus mitigation) caused by climate change will sustainable economic development no longer be possible? How would countries, where these costs reach much higher percentages of the gross domestic product than in highly developed countries, get the proper support from these?

7.3 From Rio to Kyoto

At the first Conference of the Parties (COP1) to UNFCCC in March/April 1995 in Berlin it became already clear that the anthropogenic climate change issue had grown since the Rio conference when first detection of anthropogenic climate change had been reported to be imminent in the scientific literature. Therefore, COP1 asked for a legal instrument to be ready at the end of COP3. This instrument was later named the *Kyoto Protocol*.

7.3.1 Kyoto Protocol

When the UNFCCC had been ratified by enough countries in December 1993 to become enforced in March 1994 the only "vague" commitment concerning reductions of greenhouse gas emissions by industrialized countries, the so-called Annex 1 countries of UNFCCC, was the following: aim to return in 2000 to the CO_2 emission values of 1990. Hence at COP1 the Parties asked for a further legal instrument to reduce emissions, because it was clear from climate model calculations that the soft goal of UNFCCC was far from real climate protection. At COP3 in Kyoto such a legal instrument was decided on 10 December 1997 (in reality just before noon on 11 December 1997 after a night and morning session of the negotiating body), which – for the first time – contained commitments of Annex 1 countries reaching on average 5.2 percent reduction of greenhouse gas emissions (see table 7.1) until the end of the commitment period 2008 to 2012 (averaging over 5 years was selected to avoid interannual strong variability of emissions caused for example by mild or strong winters). Besides the trend reversal of greenhouse gas emissions, which grew nearly continuously since industrialization began, the Kyoto Protocol introduces several innovations into multilateral agreements: emissions trading, joint implementation (JI), clean development mechanism (CDM) and accounting for sinks.

The Kyoto Basket, i.e. the list of long-lived greenhouse gases to be counted for emission reduction measures, together with the Greenhouse Warming Potential (GWP) with respect to carbon dioxide, needed in the accounting procedure to convert a unit mass of a gas into its CO_2 equivalent (see table 7.1). In table 7.1 greenhouse gases are ordered according to importance for the enhanced greenhouse effect. Please note that short-lived anthropogenic greenhouse gases like tropospheric ozone ($O_{3,trop}$) are also important, but not part of the Kyoto Protocol. In addition chlorofluorocarbons (CFC), also strong long-lived greenhouse gases, are regulated under the *Montreal Protocol* of the *Vienna Convention to Protect the Ozone Layer*.

Table 7.1 The Kyoto Basket

Greenhouse Gas	Chemical Symbol (abbreviation)	GWP (IPCC, 2007a) for a 100 year horizon
Carbon Dioxide	CO_2	1
Methane	CH_4	25

Greenhouse Gas	Chemical Symbol (abbreviation)	GWP (IPCC, 2007a) for a 100 year horizon
Nitrous Oxide	N_2O	298
Perfluorinated Hydrocarbons	PFCs (CF_4, C_2F_6, …)	7,390 – 12,200
Fluorinated Hydrocarbons	HFCs	124 – 14,800
Sulphur Hexafluoride	SF_6	22,800

7.4 Climate Protection Goals in Europe and Germany

The term "climate protection" is anthropocentric. We want to avoid rapid climate change to which nature, agriculture and industry cannot adapt, thereby endangering the well-being of humankind. Here climate protection is understood as the measures taken to dampen **anthropogenic** climate change. At present the European Union and some of its member countries are leading concerning climate protection goals but in parts also concerning emission reduction measures. It is difficult to assess whether even the modest goals set can be reached, because

1. the only international binding measures within the Kyoto Protocol are only set for the 2008 to 2012 period,
2. most national goals are not binding,
3. negotiations for binding measures for the post-Kyoto period have only started in 2005 and should lead – as stipulated within the Kyoto Protocol – to new measures until the 15th Conference of the Parties to UNFCCC in 2009 in Copenhagen.

Overall, i.e. when summing up emission reductions and emission increases since 1990, the Kyoto Protocol can be fulfilled, since many industrialized countries with economies in transition, the former socialist countries under the influence of the former Soviet Union and the Russian Federation itself, have gone in parts through major economic recession; and many OECD countries have taken some modest measures, decreasing their emission increase rates and some – like Germany and the United Kingdom – reducing emissions strongly due to both climate protection measures and restructuring of industry.

The next subsection will give the climate protection policy in Germany in more detail before the European approach will be presented briefly.

7.4.1 Emission Reduction Goals and Measures in Germany

When the Physical and the Meteorological Societies in Germany issued in March 1987 a joint brochure entitled "Warning of Global Climate Change Caused by Mankind", to which one of us (H. Grassl) contributed, the political reaction was: Firstly, the establishment of a "Scientific Climate Advisory Board" by the Federal Government, and secondly soon thereafter in 1988, the establishment of an Enquête Commission "Precautionary Measures to Protect the Earth's Atmosphere" by the German Parliament. Enquête Commissions are equally composed of parliamentarians and scientists or outside experts, in the above case 13 members from each group. The latter's public relation activities brought the scientific assessments of ozone depletion, tropical forest destruction and the enhanced greenhouse effect to the attention of most German citizens already in 1989. Its internal reporting led to a cabinet decision in June 1990 to reduce CO_2 emissions in Germany by 25 percent until 2005 with reference to the 1987 emissions. The goal has been reached with slightly above 20 percent only in parts, roughly half of it by political measures, the other half by integration of the former German Democratic Republic into the Federal Republic of Germany on 3 October 1990, which led – for example – to the dismantling of several large, but inefficient and polluting lignite burning power plants.

A further follow-up was the so-called feed-in law, allowing all electricity generating installations using renewable energy sources access to the electric grid at fixed prices guaranteed for 20 years since January 1991. It brought wind energy the breakthrough as a considerable part of the electric energy use by growth rates of more than tenfold in ten years (~ 26 percent growth per year).

A similar breakthrough for solar energy (mainly photovoltaics) and electricity from biomass resulted from the Renewable Energies Law of 2000 and its later modifications. At present about 13 percent of electric energy in Germany stem from renewable sources. The ranking is: wind energy, hydro power, biomass power plants and photovoltaics.

The technical innovations stimulated by this support for renewable energies has caused massive growth of export of renewable energy plants and technology, making Germany the leading country in both wind and solar energy technology in 2006.

7.4.2 Emission Reduction Goals and Measures in the European Union

The proactive role of the European Union in the climate policy debate is undisputed. However, in a world with groups of countries whose income depends nearly totally or substantially on fossil fuel exports, the pressure on Europe to strongly lead also in measures is low. As the European Union is the only area in the world where the economically stronger ones support the weaker ones over decades, burden sharing is a must, exemplified in the EU-15 agreement how to reach the Kyoto Protocol goal of minus 8 percent for the greenhouse gas emissions due to the gases listed in the so-called Kyoto basket (see section 7.3). Firstly, the minus 8 percent goal is the highest within the Kyoto commitments of industrialized countries, secondly, the burden sharing spans a wide range from minus 27 percent for Luxemburg to plus 40 percent for an economically weaker member like Portugal, which still should grow economically faster than the average member country (see table 7.2 for details).

In order to reach these goals the EU took – as the first group of countries – advantage of a new political instrument that will also be part of the Kyoto Protocol, emissions trading. Since January 2005 a European Union-wide CO_2 emissions trading scheme is in place, covering about 50 percent of all CO_2 emissions (large emitters only). It was started before the Kyoto Protocol became binding (16 February 2005). As often, when a new policy instrument is started, it suffers from several deficiencies. Some of these are: distribution of CO_2-certificates free of charge to the emitting industries, weak incentives to reduce emissions due to too high allowances for emissions, allowances set by member countries often protecting energy-intensive industries.

Table 7.2   **CO_2-reduction commitments of EU-15 member states within the Kyoto Protocol**

Member State	Target 2008-2012 under Kyoto Protocol and 'EU burden sharing' (percent)
Austria	-13.0
Belgium	-7.5
Denmark	-21.0
Finland	0
France	0
Germany	-21.0
Greece	-25.0
Ireland	-13.0
Italy	-6.5
Luxembourg	-28.0
Netherlands	-6.0
Portugal	+27.0
Spain	+15.0
Sweden	+4.0
United Kingdom	-12.5

In late 2006 the second national emission reduction allocation plans of several member countries, including Germany, have been criticized by the European Commission, mostly covering the above weaknesses. Resubmitted plans now show more stringent reductions for the coming years.

Until now greenhouse gas emissions by international ship and air traffic are not included in the Kyoto Protocol. The EU plans to integrate international air traffic into the CO_2 emission reduction policy.

7.4.3 Reduction Goals

From scientific calculations it became clear that the maximum tolerable mean global warming of 2°C within the 21st century translates into at least halfing global emissions until 2050, which again translates into 80 percent emission reduction for industrialized countries until this date. Since the EU has set the 2°C goal, the Kyoto commitments can only be a start. At COP 12 of UNFCCC in Nairobi in December 2006 negotiations have led to the following tentative goals, only expressed under conditions of similar reactions by other parties:

– 40 percent for Germany until 2020

– 30 percent for the European Union until 2020

On 9 March 2007 the European Union Council decided to at least set the following binding targets until 2020: – 20 percent CO_2-emissions, 20 percent energy efficiency gain, 20 percent of primary energy from renewable sources. The key questions, however, in view of the global challenge of climate change are:

How can emerging countries be integrated into the Kyoto follow-on process, as their development path is decisive for future emissions? What incentives are needed that their energy supply system uses rapidly renewable energy resources.

Answering these questions depends also strongly on the EU climate policies and technological innovation in developed countries.

7.5 Sustainable Development Strategy (of the EU)

It is clear that the present structures of economies worldwide will not lead to a sustainable development. We deplete non-renewable resources at high rates without engaging to the extent necessary for their replacement by renewable resources. Because rapid climate change can destroy all efforts to reach sustainable development climate change mitigation is the most urgent measure to keep the corridor to sustainable development open. Climate change mitigation encompasses not only CO_2 emission reduction, but many more long- and short-lived greenhouse gases and its precursors must be reduced, for example nitrogen oxides ($NO + NO_2$) that – jointly with hydrocarbons – lead to tropospheric ozone, whose climate change contribution is already at the same level as accumulated methane. A sustainable development strategy – as adopted by the EU – has to deal with ecologic, economic, social and cultural goals going well beyond mere climate change mitigation and adaptation measures. It is a continuous process searching for the best approach to remain in a sustainable corridor by joint action of science, politics and society in regions, countries and at the international level. As expressed in Schellnhuber (2001), it is a co-evolution involving all actors from the beginning. We still have to learn this joint action.

8 Evaluation of Climate Effects

8.1 The General Evaluation Problem

The reduction of negative climate change impacts which influence every human activity, in private life as well as in business, is indeed urgently needed. For the improvement of the current situation a broad variety of options is available, or will be developed to fill current gaps. However having such options defined another problem arises. In a concrete situation we have to choose one of the options, and only one, for a definite solution. Which option is to be selected? This is not a simple question and the decision making becomes a complex task.

Let us look at various situations: Some of the solutions proposed by the systems analysis efforts may be of such a character that they lead in the right direction especially if they have extra positive side effects unrelated to climate. Such an example is the reduction of private transport in a city in favour of public transport if it is available. This would result in reduced emissions of carbon dioxide and other greenhouse gases, as well as toxic gases from the combustion of fuel. Traffic noise pollution would be reduced. There would be lower risks of traffic accidents as well as health benefits. Moreover reduced use of fossil fuels lowers its overall consumption.

The above example is not typical of the decision making situation when reducing greenhouse gas burdens: In most cases a certain measure will result in different effects which can be both positive and negative.

Some examples of decision problems related to climate change are given in table 8.1. Details are discussed in the following chapters.

Table 8.1   Interdependence of climate related problem solutions

Problem solution in question	Possible effects
How can financial resources be invested most efficiently for climate protection?	Most measures for climate protection cost money. The more a technical solution costs the better normally the effect is. However the money spent could also be used for other kinds of improvements which may result in better climate protection. The best solution has the highest benefit/cost ratio.

Problem solution in question	Possible effects
	The idea is behind CDM (Clean Development Mechanism) on an international scale where expensive measures in one country are substituted by cheaper measures in other countries which result in a larger climate effect for the same cost.
Is the reduction of climate gas emissions more important than the reduction of emissions of cancer causing substances?	Climate gas reduction is influential on a global scale. Cancer causing substances are important factors for human health. Both cannot be compared directly. A decision on the side effects for human beings is necessary.
Under which conditions is the substitution of fossil fuels by biofuels efficient with respect to climate?	Fuels can be produced from fossil or from renewable energy sources. Renewables are generally preferred. However they are produced by agriculture, and become competitive with human nutrition. Need for land may foster deforestation and stress biodiversity. Production needs would influence input of fertilizers and pesticides which may negatively affect soil and groundwater, and cause air pollution. The production of fertilizers and pesticides consumes fossil fuel. Environmental effects of biofuel production are to be considered.
Should waste be deposited with or without pre-treatment?	Waste pre-treatment reduces the organic matter and separates recyclable materials. Thus methane production in the landfill is lowered. Virgin raw matter as well as production efforts for production of materials can be reduced, thus reducing climate gas emissions in industrial production. Long term landfill effects of liquid leachates and emissions are reduced. But low methane emissions from landfill also means that there is no possibility of collecting gas in a gas recovery system and using it as an alternative energy source instead of fossil fuels.
Is the reduction of emissions of nitrous oxide (N_2O) from a chemical process more climatically effective than the reduction of emissions of CO_2?	Both substances are greenhouse gases. But their GWP is different: the reduction potential of one unit in N_2O is 298 times larger than one unit in CO_2. Such effects can be clearly calculated. However, the steps to reduce both gases are different and must be taken into consideration.

Sustainable Development Strategy (of the EU)

Problem solution in question	Possible effects
Is it environmentally effective to reduce greenhouse gas emissions with the use of end of pipe technologies?	End of pipe technologies, such as washers, biofilters, or gas incineration are able to reduce direct CO_2 emissions. But this causes extra equipment and may be accompanied by higher energy input for the process which cause indirect climate effects as well as other environmental burdens. Sometimes only environmental effects are transferred from one media to another (e.g. from air to water).

Table 8.1 indicates that a clear decision for a certain option under discussion is only possible if decision criteria are available which focus on climate effects, and at the same time consider other environmental as well as economic and social aspects.

This problem is not new, and it is not only true for climate protection. Therefore a variety of evaluation procedures (or assessment tools) has been established, which can be used for such decision tasks. They normally focus on special items or specified groups. A general decision procedure which simultaneously covers all aspects of sustainability does not yet exist. However, climate effects play an important role in most of the procedures. A graphic representation of some assessment tools is given in figure 8.1.

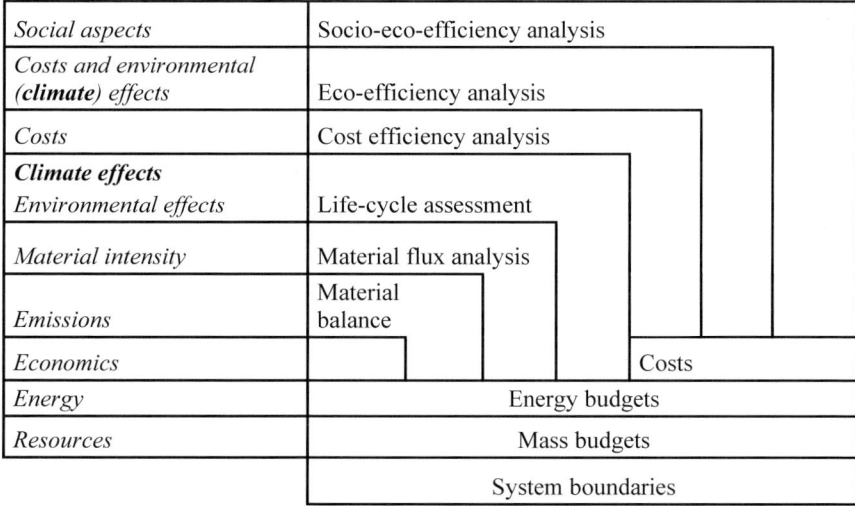

Figure 8.1 **Hierarchy of assessment tools (Gohlke, 2006)**

Obviously all methods are based on matter and energy budgets within certain systems boundaries which are to be defined as a pre-condition for a sensible evaluation process. Economic evaluation is based on costs only. Environmental effects are solely considered in the life cycle assessment. Combinations of economic as well as ecological factors may be assessed using methods such as eco-efficiency or cost-efficiency analysis. The trio of economy, environment, and social aspects is evaluated by the socio-eco-efficiency analysis (BASF, 2004). This method, as well as others of this kind, is relatively close to the needs of a sustainability oriented evaluation criterion but focuses on selected aspects of the item under consideration.

A more detailed description of some practical assessment tools is given in table 8.2.

Table 8.2   Selected environmental assessment tools

Assessment tool	Purpose of application	Example
Environmental audit	Actual ecological performance is analysed, and targets are set for future environmental performance on company levels	Auditing of a production plant, including situation report and target definition
Environmental Impact Assessment (EIA)	Analysis of the environmental impact potentials of a planned facility or other activities	Choosing a production site for a chemical plant or a landfill site
Life-Cycle Assessment (LCA)	Comparison of the environmental effects of the whole life cycle of products, processes, services, or other activities with the same function; identification of improvement and optimisation potentials	Comparison of beverage bottle systems Identification of key process steps in climate control Comparison of the climate change potential of different substances Comparison of climate and health impacts
Risk Assessment	Estimation of impacts of an event and its probability	Assessment of the risks of landfill gas emissions after destruction of a landfill cover

Assessment tool	Purpose of application	Example
Substance and material flow management (Bringezu, 2000; Brunner, 2004)	Balancing of the material flows in an observation unit (e.g. company, regional, national level) Identification of flows of dangerous substances and of causes of environmental problems in a region Control of flows according to given criteria	Analysis of wood balance in a region to find the best use of it as a renewable resource
Sustainable Process Index Assessment (Sandholzer, 2005)	Definition of the impacts of a certain process in terms of the area needed per process unit	Comparison of ethanol production on the basis of renewable or fossil energy sources
Eco-efficiency assessment	Combined assessment of the impacts of a process, a technology, or a service on ecological (LCA) as well as economic criteria	Definition of the best waste management technology for a unit of waste, including deposition, pre-treatment, and waste combustion for a region

Actually used for the estimation of climate effects of processes and services is Life Cycle Assessment (LCA). It is therefore explained in more detail in the following chapter.

8.2 Life Cycle Assessment for Climate Control

8.2.1 Background issues

Life Cycle Assessment (LCA) compiles the potential environmental aspects throughout a product's life, including raw material acquisition, production, use of the produced good or service, and its disposal. Life cycle assessment is often thus termed "from cradle to grave". It serves:

- to compare environmental effects of different products
- to identify environmental key issues, such as climate effects
- to improve and to optimize products, the production thereof as well as their use.

The term "product" refers not only to a material, but includes also processes and services.

The definition as well as the whole procedure are standardised by ISO 14040 norm group (CEN, 1999). This norm provides a framework and defines the key methodological needs, to make LCAs comparable with each other, independent of the institutions, or country, where the LCA analysis was performed.

There are several activities worldwide to improve the LCA methodology and support its application:

- On the level of the United Nations Environmental Programme (UNEP) a Life Cycle Initiative was established as a response to the call from governments for a life cycle economy in the *Malmö Declaration* (in 2000). It contributes to the 10-year framework of programmes to promote sustainable consumption and production patterns, as requested at the *World Summit on Sustainable Development* (WSSD) in Johannesburg (in 2002). This initiative develops and disseminates practical tools for evaluating the opportunities, risks, and trade-offs associated with products and services over their whole life cycle (UNEP, 2007).

- The European Commission (EC) launched a simplified method which condenses the basic methodology of ISO 14040 into an EXCEL calculation spreadsheet which already contains basic data on environmental impacts. The aim is to enhance the application of LCA activities in small enterprises for the environmental and climate related improvement of their products or services (VHK, 2005; UNEP, 2007).

- The Society of Environmental Toxicology and Chemistry (SETAC) provides an international infrastructure to support Life Cycle Assessment groups in advancing science, practice, and application of LCA and related approaches worldwide. The organization serves as a focal point of a broad-based forum for the identification, resolution, and communication of issues regarding LCAs, and facilitates, coordinates, and provides guidance for the development and implementation of LCAs in close cooperation with each other. Core tasks include planning and organizing LCA sessions and conferences (such as the Annual Meetings in Europe and North America and the LCA Case Studies Symposium), coordination of topical working groups, preparation and global integration of LCA publications, and promotion of the UNEP/SETAC Life Cycle Initiative (UNEP, 2007).

- Several countries such as Denmark and the Netherlands also have more detailed guidelines prepared for internal use, including databases for LCA activities. In Germany several calculation programmes are available free of charge in LCA procedures (e.g. GEMIS, Öko-Institut, 2006).

- One of the largest repositories of corporate greenhouse gas emissions data in the world is the Carbon Disclosure Project (CDP) which as a non-profit organisation is acting to facilitate a dialogue, supported by quality information, from which a rational response to climate change will emerge (CDP, 2007).

8.2.2 LCA Methodology

LCA after the general ISO framework consists of four phases (CEN, 1999):

- In the *Goal and Scope Definition* the purpose of the study and the boundary conditions are defined.
- In the *Life Cycle Inventory* phase emission and resource data of the process under study are gathered.
- In the *Life Cycle Impact Assessment* phase potential environmental impacts of the emissions as well as resource consumption facts gathered are analysed and quantified.
- In the *Interpretation* phase the results are interpreted by a group consisting of the LCA team, the client, as well as of independent people interested in the topic, and conclusions are drawn.

A more detailed description of the items to be worked out in the four phases and the questions follows.

8.2.2.1 Step 1: Definition of goal and scope

In this first LCA phase the purpose and the boundary conditions of the study are defined. Amongst the issues to be described the definition of the "functional unit" which captures the functions of the study is most important, since it provides the reference to which input and output are related. Some examples of functional units for climate related studies are given in table 8.3.

Table 8.3   Examples of functional units for climate related studies

Example	Typical functional unit
Comparison of transportation systems	Transport of one ton of raw material over a distance of one kilometer
Comparison of types of product containers for chemical substances	One container carrying one ton of material
Comparison of waste treatment practices	Treatment of a certain amount of waste, e.g. one ton of municipal solid waste

In principle, all processes from "cradle to grave" must be included in such a study. But in practice all those processes which are identical for the products compared and thus do not influence the result, such as production efforts for machinery, can be eliminated from the scope of study.

The following example of the definition of the scope for a life cycle assessment demonstrates an approach for biofuels – see table 8.4; of details the technologies see chapters 10 to 13. All technological measures dealing with manufacturing of machinery and infrastructure are excluded since these items are the same for all processes. However, there is a choice whether or not to include the production of the feedstock of the biofuel process, like agricultural inputs or raw material for processing. This depends on the aim of the analysis.

Table 8.4   Items to be considered for biofuel production LCA (BR&D, 2005)

To be included	To be excluded
Origination of the raw material, such as production, mining, or extraction.	Manufacture of machines and physical infrastructure used in product or raw material origination, manufacturing, distribution, use, and disposal.
Manufacture or processing of the raw materials into the finished product.	
Manufacture of material inputs consumed in origination, manufacturing, or processing.	Manufacture of machines used to manufacture materials consumed in the origination, manufacturing, distribution, use, and disposal processes.
Transportation of raw materials and finished products to points of use or sale.	Operation of ancillary offices and activities such as business travel, not directly associated with the origination, manufacturing, distribution, use, and disposal of the product.
Product use, including combustion, if applicable.	
Disposal of waste products from manufacture and use.	

8.2.2.2   Step 2: Life-cycle inventory analysis

The inventory analysis involves data gathering and calculation procedures to quantify inputs and outputs, i.e. resources and emissions. This is based on setting up flowcharts of the system under study, which contain all details of the processes and the interconnections of process steps. A definition may be necessary at this stage of the work which mass and energy fluxes are important and which eventually can be eliminated to reduce the complexity of the LCA. Indeed, a pre-condition for such a definition is knowledge about the process effects. These may be a result of previous steps in a trial and error procedure, which is typical for LCA measures.

Data gathering can be a time consuming task. Potential data sources may be primary data from process studies, or from literature, or expert judgement.

Life Cycle Assessment for Climate Control   89

Also public databases, such as the GEMIS database (Öko-Institut, 2006) can be used, which offer inventory data for a large number of processes if, as is often the case, actual primary data cannot be extracted.

The data have to be normalised to the functional unit.

8.2.2.3 Step 3: Life Cycle Impact Assessment

This step of the Life Cycle Assessment aims to evaluate the magnitude and the significance of the potential environmental impacts of the system under study. It involves three mandatory elements: Most important is the selection of i) impact categories, ii) indicators for these categories, and iii) models to quantify the contribution of resources and emissions to it. This also may comprise a ranking of the indicators. In the classification step, the inventory data have to be assigned to the impact categories. Afterwards, in the characterization procedure, the contribution of the inventory data has to be quantified for the chosen impacts.

In life cycle assessment practice the following impact categories with specific indicators were established (see table 8.5). Table 8.5 also indicates some examples of substances which primarily influence these impact categories. Besides impact categories given in table 8.5, which are used in the LCA after ISO 14040, other criteria are used for environmental analysis, e.g. energy and material needs per unit of product, which may also have a direct or indirect effect on climate.

Table 8.5   LCA impact categories and indicators

Impact category	Indicator	Description and climate relevance	Substances
Global warming potential, GWP	CO_2	Contribution to global warming by its heat absorption capacity. Value depends on the time horizon considered. Typically a hundred years time horizon is used. High climate relevance.	CO_2, CH_4, N_2O, SF_6, HFCs, PFCs
Ozone depletion potential, ODP	R11 (CCl_3F)	Contribution to the depletion of the stratospheric ozone layer by persistent chlorine and bromine hydrocarbons. Global effects on biosphere by UV radiation, effects on human health. Also climatically relevant.	ODSs (see chapter 9.), HCFCs

Impact category	Indicator	Description and climate relevance	Substances
Photochemical ozone formation potential, PCOP (Summer smog potential)	Ethylene	Contribution to the formation of oxidizing substances, e.g. ozone, mostly by reactions between NO_x and NMVOCs under the influence of UV-radiation in the troposphere. Effects on human health and ecosphere (e.g. damage of forests). No direct, but indirect climate effects.	NMVOCs, CH_4
Acidification potential, AP	SO_2	Contribution to the acidification of an environment. Regional effects. No direct climate relevance, but strong indirect relevance after aerosol particle formation.	SO_2, NO_x, NH_4, HF, HCl
Nutrition potential, NP	PO_4^{3-}	Contribution to the production of biomass. Regional effects. No direct climate relevance.	NO_x, NH_3, NH_4^-
Cancer potential (short term and long term)	"Risk units"	Contribution to cancer risk caused by certain substances after a contact with it. Short and long term effects. No environmental effect, no climate effect.	Cd, Hg, Cr_{VI}

To define how much a process contributes to the impact categories, the effects of all material and energy fluxes of the relevant process steps and their individual impact have to be considered. The total of the impact category is given as the sum of the effects of all substances "k" in all process steps.

In the case of the GWP of a process this is given by the following equation:

$$GWP_{total} = \sum (\beta_k * GWP_{100; k})$$

GWP_{total}: total global warming potential

$GWP_{100, k}$: individual GWP of the substance k for a 100 years's time horizon

β_k: mass of emission of substance k

An example for the calculation of the greenhouse gas potentials for two variants of a simple fictitious process is given in table 8.6 using individual GWPs of the substances k (see table 7.1; IPCC, 2007a).

Life Cycle Assessment for Climate Control

Table 8.6   Greenhouse potential of two processes (fictitious numbers)

Substance k	GWP_k	Emission β_k (units/t)		$GWP_k * \beta_k$ (CO_2-eq./t)	
		Variant 1	Variant 2	Variant 1	Variant 2
Nitrous oxide (N_2O)	298	12	10	3,576	2,980
Methane (CH_4)	25	15	12	375	300
Monochlormethane (CH_3Cl)	13	0.8	1	10.4	13
Tetrachlormethane (CCl_4)	1,400	0.045	0.060	63	84
Carbon dioxide (CO_2)	1	100	200	100	200
Sum				4,124.4	3,577

The process may be characterised by emissions of five substances (see both columns 3 and 4 for the two variants of the process considered) with different individual GWPs (see column 2). The resulting GWP value for each substance is given in columns 5 and 6 for the variants 1 and 2, respectively. The total GWPs will result if the values in columns 5 and 6 are summed up. In the case given, variant 2 has a lower GWP compared to variant 1. Thus variant 2 would be chosen if the decision is made only on the basis of GWP.

It should be mentioned that this result is not a simple one: If only emission amounts were considered there would also have been preferences for variant 1, since there are lower emission fluxes for monochlormethane (CH_3Cl), tetrachlormethane (CCl_4), and carbon dioxide (CO_2). Therefore, only a combined approach leads to valide information.

8.2.2.4   Step 4: Interpretation

To reach conclusions and recommendations consistent with goal and scope of the LCA an interpretation phase is necessary where the results of the inventory analysis and the impact assessment are combined. This phase is comprised of i) the identification of the significant issues, ii) the evaluation of completeness of data, sensitivity, and consistency, as well as iii) conclusions and recommendations.

8.2.3 LCA case study: Comparison of climate effects of integrated waste management systems

Examples for LCAs are given in the next chapters for certain defined industrial, agricultural, energetic, and other processes. They focus on climate effects of the process, mainly on the greenhouse gas potential, as "mass unit CO_2-eq. per mass unit of product". As was explained the LCA methodology comprises other environmental burdens as well, which have to be kept in mind for a comprehensive decision. Therefore the following example elucidates the methodology of a full LCA. All data are after Soyez (2001). The example refers to waste management technologies which are described in detail in chapter 12.

8.2.3.1 Definition of goal and scope

The goal of the study is to find the best solution for the treatment of municipal solid waste in respect of the impact categories. This implies

- Comparison of the impacts of different waste management options
- Identification of key factors and critical paths
- Assessment of the importance of political, economic and technical background conditions
- Assessment of the ecological benefits through material flow specific waste management
- Identification of most effective improvement strategies and measures
- Systematic analysis of uncertainties
- Determination of best values in (technical) compromise situations.

The systems boundaries include the pre-treatment facilities and the landfill. Waste collection is not considered. Credits are given to recovered materials and energy resources. The consumption of materials and energy for the operation of the pre-treatment plants and the landfills are taken into account.

Since the technologies can be set up anywhere in the world, the analysis in principle is independent of any particular country. However, for a concrete solution, German conditions are used.

Reference point of the calculation is the specific contribution to the total effects of waste management in Germany, i.e. to which percentage the waste management system will contribute to the total national impact of every category if the total amount of residual waste is treated by the considered system.

Life Cycle Assessment for Climate Control

8.2.3.2 Technology description and functional unit

For waste treatment several technologies can be applied. Under EU conditions a so-called pre-treatment is necessary before deposition of waste can take place. In principle, all technologies after figure 8.2 can be applied.

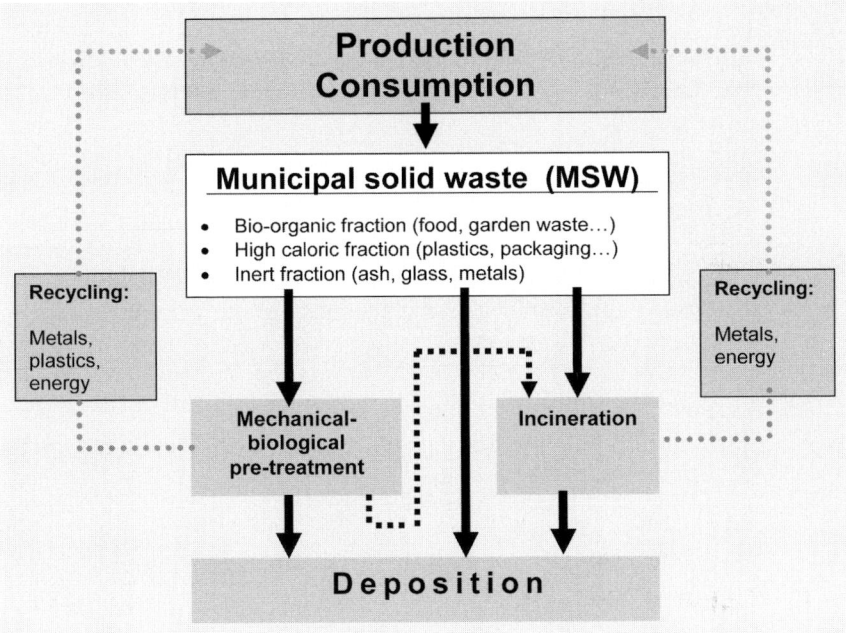

Figure 8.2 Waste management pre-treatment technology options

In the study, special waste management processes after table 8.7 are included. Technological details are described in chapter 12.

Table 8.7 Technologies included in the LCA

Technological option	Assumptions and specialities
Landfill without pre-treatment	Technology considered for comparison only. Not at the accepted state of the art in EU. No decomposition of organic matter before the landfill. Capturing of landfill emissions and energy production from landfill gas recovery.

Technological option	Assumptions and specialities
Mechanical biological pre-treatment (MBP)	High decomposition of organic matter by intensive processing ("rotting") (e.g. 8 weeks with forced aeration) and subsequent extensive decomposition. Recovery of the metal fraction. Collection and cleaning of waste gas of the pre-treatment by a biofilter system.
Combustion	Conventional combustion plant with grate firing.

The principal function of all technologies is the treatment of waste. Therefore, as the functional unit, the treatment of one ton of waste material is the yardstick.

8.2.3.3 Impact assessment results

The impact assessment is performed using the impact categories in table 8.5. Figure 8.3 displays the results of the assessment given as contributions of selected categories to the national (in this special case: German) emission inventory. Negative values mean benefits for the environment and positive values indicate an extra burden.

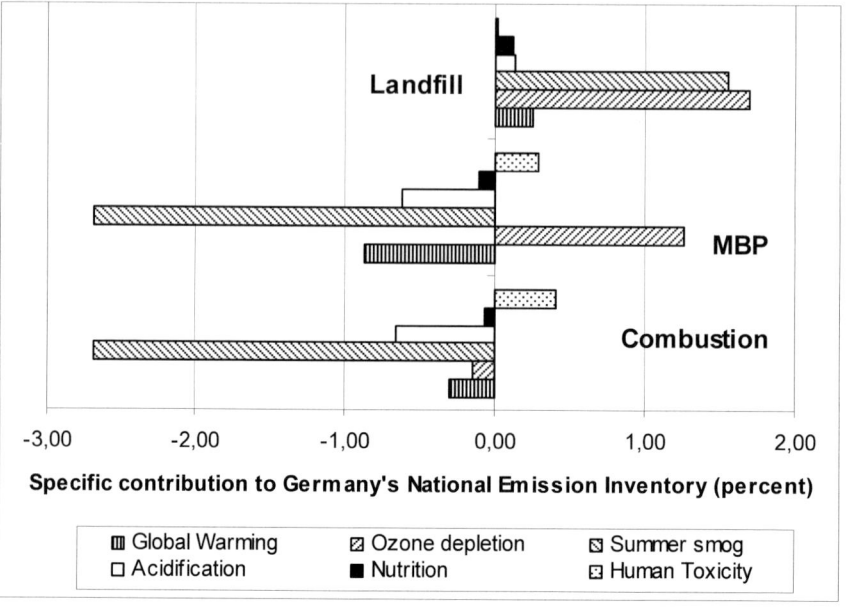

Figure 8.3   **LCA results of waste management technology options**

Life Cycle Assessment for Climate Control

8.2.3.4 Interpretation and conclusions

Figure 8.3 indicates, in terms of the impact categories, the environmental benefits and disadvantages of the waste management options. If a decision is to be made only on the base of one category, a clear decision would be possible in the case of summer smog potential as well as for climate impacts: Combustion and MBP are preferable options. For climate impacts alone MBP would be the best solution. Only direct deposition results in environmental burdens in every impact category considered. The other options result in both burdens and benefits. Therefore none of the waste treatment options is best in every criterion.

The Global Warming Potential (GWP) is mainly caused by CO_2 and CH_4. The burdens result from the energy consumption and the landfill gas emissions. However, these burdens can be equalised by credits through recovery of waste fractions for recycling or power production, so that MBP and combustion end up with a climate benefit.

The ozone depletion (ODP) and the summer smog potentials (POCP) of the landfill and the MBP are caused mainly by CFCs (CFC-11, CFC-12) and by highly volatile chlorinated hydrocarbons which are emitted during pre-treatment and in the landfill. The specific contributions of the burdens exceed one percent. During combustion, these substances are almost completely destroyed, so that no burdens result from them in this technology. On the other hand, in the combustion facility, energy is recovered, which results in benefits. For MBP benefits from material and energetic recovery prevail.

The acidification potential (AP) is caused by SO_2, NO_x, and ammonia emissions, the nutrification potential (NP) by ammonia and NO_x emissions. Both burdens are comparatively low and are balanced through credits for material and energy recovery.

The Human Toxicity Potential is predominated by the gaseous emissions of heavy metals such as chrome, cadmium, and nickel. These metals are mobilised to a high degree when the waste or waste fraction are combusted. This option, as well as MBP option which includes the recovery of a RDF-fraction, end up with higher burdens. In the landfill a certain percentage of heavy metals can be considered to be stored over a period of several thousand years, depending on the buffer capacity and the creation of humus-like substances within in the landfill. Therefore the landfill option has the lowest Human Toxicity Potential.

This chapter explained only the methodology of LCA and its application for climate control measures; details of the waste management processes considered as well as more examples of greenhouse gas effect evaluations for processes of various kinds are given in the following chapters.

9 Climate Effects and Mitigation Potentials of Economic Sectors

Economic sectors comprise industrial manufacturing, agricultural production, energy production, and waste management processes. Every such process results not only in the products wanted, but also in by-products, wastes, and emissions. Amongst the emissions, typically, greenhouse gases occur. Hence, every process is climate relevant.

The climate effect of a single process is the result of the emission amount and the specific GWP of the emissions and can be calculated as the product of mass and individual GWPs (see table 8.6). It depends on the specific process and the conditions of its application. Thus, the choice of the raw material, the technology, the apparatus, the cleaning device, and others may influence the climate impacts of the production process. An optimisation of the process with reference to a decision criterion reflecting the climate effects is thus a necessary precondition for the choice of the best technology (see chapter 8.1).

To get a first impression of the importance of the different production sectors for climate change, a comparison of the total GHG fluxes in a country or in a region seems useful. The current situation of climate gas emissions by the most important six substances (*Kyoto gases*) as well as their contribution to the sectors of the whole world's economy and some selected regions (Germany, EU, USA) is given in table 9.1.

Table 9.1   Kyoto gas emissions in selected regions

	Country/region			
	Germany (UBA, 2007)	EU-15 (EC, 2007)	USA (EPA, 2007b)	World (IPCC, 2007c)
Reference year	2005	2005	2005	2000
Total emission (Mio t CO_2-eq.)	1,002	4,192	7,260	43,000
Contribution of Kyoto gases	**Percent of total GHG emissions**			
CO_2	87.1	83.1	83.9	77

Contribution of Kyoto gases	Percent of total GHG emissions			
CH_4	4.8	7.4	7.4	15
N_2O	6.6	8.0	6.5	8
HFCs	0.9	1.3		
PFCs	0.1	0.1	2.2	1
SF_6	0.5	0.2		

Figure 9.1 represents to which extent the economic sectors contribute to the total GHG emission of selected contries or regions and world's mean values.

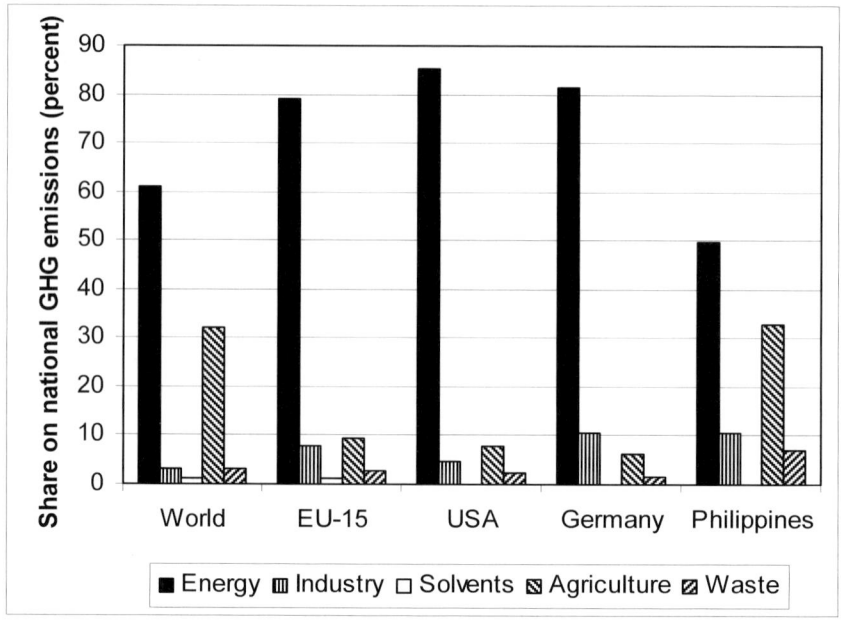

Figure 9.1　GHG emissions economic sectors in selected economies

Obviously, in industrialized countries, the energy sector contributes to about 80 percent or more, followed by agriculture and industrial production with around 5 to 10 percent each. Waste treatment with only one to three percent is more or less marginal.

Activities in climate improvements therefore should start with improvements in the energy sector, but every percent of reduction also in the other sectors counts. Therefore, a broad range of activities is necessary.

9.1 Processes and Typical Emissions

With respect to the emission types, CO_2, N_2O, CH_4, and so called high global warming potential gases (high GWP gases, such as PFCs and HFCs) dominate. GWPs are given in table 7.1 (see chapter 7) for a 100-year time horizon.

In which process nitrous oxide, methane, or high GWP gases predominate is given in table 9.2. CO_2 which is the most important emission (by more than 90 percent of total emission, see table 9.1) is not explicitly mentioned here, since it is relevant in practically every process.

Table 9.2   **Emissions of typical production processes**

	Methane	**Nitrous oxide**	**High GWP Gases**
Energy production	Coal mining; Natural gas and oil systems; Stationary and mobile combustion; Biomass combustion	Biomass combustion; Stationary and mobile combustion	
Agriculture	Livestock manure management; Livestock enteric fermentation; Rice cultivation; Agriculture residue burning; Prescribed burning of savanna; Agricultural soils	Livestock manure management; Agricultural soils; Agriculture residue burning; Precsribed burning of savanna	
Waste management	Landfills; Waste combustion; Mechanical-biological waste treatment, Composting; Wastewater treatment	Human sewage; Waste combustion	

	Methane	Nitrous oxide	High GWP Gases
Industrial processes	Petrochemical production; Silicon carbide production; Iron and steel production	Adipic and nitric acid production; Caprolactam production; Solvent use; Fugitives from oil and natural gas systems, fugitives from solid fuels	Substitutes for ozone depleting substances (HFCs, PFCs); Production/manufacturing of HCFC-22 (HFC-23); Primary aluminum (PFCs); Magnesium (SF_6); Production of electrical equipment (SF_6) and semiconductors (PFCs, SF_6, HFCs)

9.2 General Mitigation Potentials

Mitigation means reducing GHG emissions by technical means, of every sector including industrial processes, agriculture, waste management, and energy.

A broad variety of measures is currently available; others are under intensive research and will become practicable in the shorter or longer term. This includes organisational practices such as energy auditing, the application of management systems, GHG inventory and reporting systems, or benchmarking, which are similar for every process considered. Other practices are sector oriented and refer to technologies or processes.

In table 9.3 a survey of such practices including an estimation to which extend they would affect CO_2 reduction is given; the emissions reduction potential is estimated for the year 2030. Details of special practices and measures are included in the following chapters.

Table 9.3   Current and future mitigation practices and effects (Vattenfall, 2007)

Sector	Mitigation practice	Estimated effect (Mio t CO_2-eq./a)
Energy	Carbon Capture and storage (CCS, see 9.4)	3,100
	CO_2-efficient fuel (use of natural gas instead of coal)	400
	Use of renewable energy (wind, solar, biomass)	1,500
	Reduced demand (increased end use efficiency)	3,700
Industry	Improved electric motor drive systems	1,200

General Mitigation Potentials 101

Sector	Mitigation practice	Estimated effect (Mio t CO_2-eq./a)
	Smelt reduction in steel making	200
	Increased industrial energy efficiency	400
	Substitution of fossil fuels in industry	300
	Feedstock substitution in cement and clinker production	700
Transport	Fuel efficient vehicles (vehicle body improvement, hybridization)	1,500
	Fuel switch (biofuel, flexi-fuel)	400
	Demand reduction (reduced travelling distance, public transport, intelligent traffic management)	400
Building sector	Lighting (energy efficient lamps)	400
	Effective white goods (e.g. refrigerator, washing machine)	200
	Reduced stand-by losses	200
	Water heating and air conditioning (use of efficient compressors, heat pumps)	500
	Improved building insulation (wall, window, roof)	1,700
Forestry	Reduced deforestation	3,300
	Increased forestation	3,500
Agri-culture	Reduced flooding of rice fields (dry cultivation, slow release fertilizers)	100
	Improved animal handling	200
	Shift in fertilizers (optimized application, nitrification inhibition)	700
Waste management	Recycling	300
	Capturing and use of landfill gas	400

Though only the figures given in table 9.3 are rough estimates it becomes evident that considerable effects can be achieved after application of the mitigation activities. The sum amounts nearly 27,000 Mio t CO_2-eq. (to be reached in 2030).

Some of the practices in table 9.3 are available at low or even no cost, but in general increased capital investment or operating costs must be considered for mitigation activities. Therefore the input of money affects the mitigation results which can be achieved in practice.

The estimated overall mitigation potential on a global scale in dependence of the cut-off rate for marginal abatement costs ("carbon price") is presented in figure 9.2.

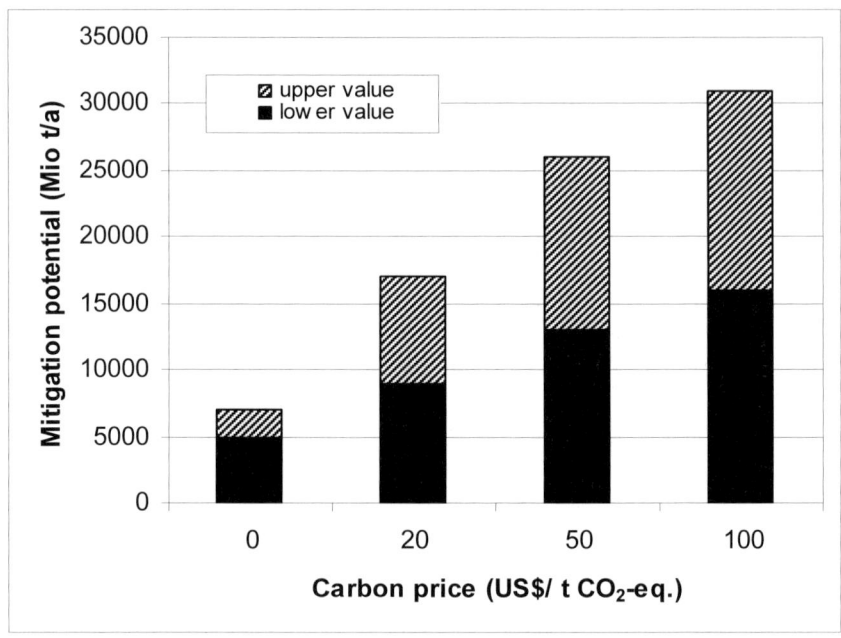

Figure 9.2 Global mitigation potential for carbon price categories (IPCC, 2007c)

Figure 9.2 indicates that a global mitigation potential of about 6,000 Mio t CO_2-eq./a is available without extra costs. For mitigation in the range of 9,000 to 31,000 Mio t CO_2-eq. a carbon price between 20 and 100 US$ per ton CO_2-eq. is expected depending on the scenario considered. This is two-thirds of the recent total of world's GHG emissions (43,000 Mio t CO_2-eq./a – see table 9.1). Its application therefore would reduce the GHG emissions to an acceptable low level.

The term "carbon price" is sometimes represented by the shadow price of an additional unit of CO_2 emitted or by the rate of carbon tax or the price of emission permit allowances (IPCC, 2007d). For a company it is more profitable to invest into technology instead of paying the carbon tax. As an example, according to figure 9.2, at a carbon price of 20 US$/t CO_2-eq. an average abatement effect between 9,000 and 17,000 Mio t CO_2-eq./a can be expected in the long run on a global scale.

9.3 Mitigation Potential by Carbon Capture and Storage (CCS)

The largest mitigation potential among the known technologies is addressed to Carbon Capture and Storage (CCS) which is not sector specific. The abatement potential is estimated to be 3,100 Mio t CO_2/a (Vattenfall, 2007).

CCS comprises i) the generation of a gas stream with a high concentration of CO_2 and capture it from industry, energy related and other sources, ii) its transportation to a storage location, and iii) long-term isolation from the atmosphere either in suitable geological formations, such as oil and gas fields, inminable coal beds and deep saline formations, or in the oceans, e.g. by direct release into the ocean water column or onto the deep seafloor.

CCS may be applied for fossil CO_2 emissions, but also for CO_2 emissions from renewable sources, which would result in an extra benefit for climate. It can be applied to large point sources, i.e. with capacities of more than 0.1 Mio t CO_2 per year. These include fossil fuel or biomass energy facilities, major CO_2-emitting industries, natural gas production, or synthetic fuel plants. Worldwide, such large stationary emit nearly 13,500 Mio t CO_2-eq. per year (see table 9.4), which is about one third of world's total GHG emissions.

Table 9.4   Point CO_2 emission sources (IPCC, 2005)

Process	Number of sources	Emissions (Mio t CO_2-eq./year)
Fossil fuel		
Power	4,942	10,539
Cement production	1,175	932
Refineries	638	798
Iron and steel industry	269	646
Petrochemical industry	470	379
Oil and gas processing	Not available	50
Other sources	90	33
Biomass		
Bioethanol and bioenergy	303	91
Total	7,887	13,466

Net reduction of emissions through CCS depends mainly on the fraction of CO_2 captured. Available technologies capture up to 95 percent of the CO_2 processed in a plant. However, there are losses in the overall efficiency of the process due to extra requirements of energy, transport and storage, which result in increased CO_2 emissions for the same amount of end product. A power plant with a CCS system needs 10 to 40 percent more energy compared to a reference plant without CCS. Therefore the final result is reduced CO_2 emissions of about 80 to 90 percent, depending on the type of the capture system (see Figure 9.3).

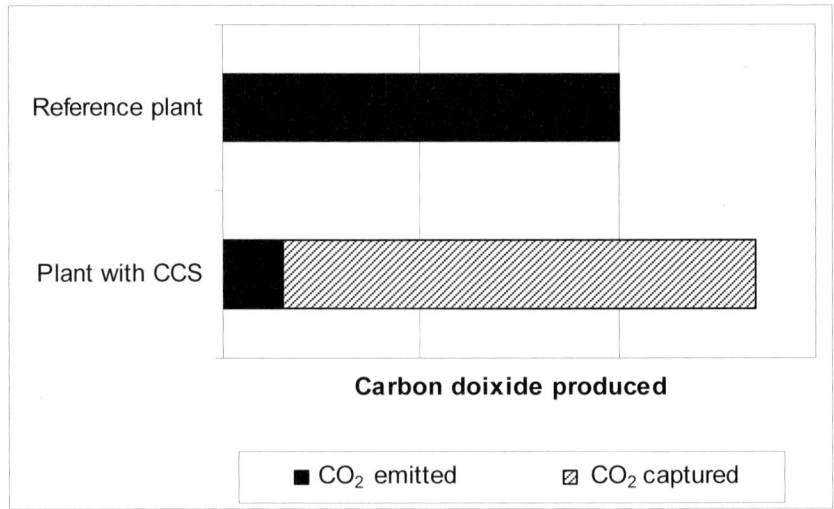

Figure 9.3   Net CO_2 emission after CCS application

In most cases, storage is not possible on the site of the production facility, so that transportation is necessary. Pipeline transport is preferred for a distance up to 1,000 km. For amounts smaller than a range of few Mio t CO_2 annually or for larger distances overseas the use of ships is commercially attractive. For both options approved technology and infrastructure exist. Pipeline transport of CO_2 is practiced e.g. in the USA where more than 40 Mio t CO_2 annually are transported in a pipeline system which is more than 2,500 km in total. Shipping is comparable with sea transportation of liquefied petroleum gas.

Storage of CO_2 in geological formations uses many of the same technologies that are applied by the gas and oil industry. Combinations of CCS and enhanced oil recovery (EOR) from boreholes are applied and result in enhanced revenues. Total storage capacity is in the range of 2,000 Gt CO_2. Fractions retained in geological reservoirs are estimated to very likely exceed 99 percent in appropriately selected and managed plants in a hundred years period; the same amount is likely to be retained after 1000 years.

In oceans CO_2 is stored either by injection and dissolving the gas into the water column below 1,000 m via a pipeline or ships, or by depositing via a fixed pipeline or an offshore platform at depths lower than 3,000 m. CO_2 in this environment is denser than water so that long term storage is possible. Storage capacity is in the range of thousands of Gigatons. Technological and environmental impacts are currently being researched.

Energy and economic models indicate that CCS will mainly contribute to CO_2 emission reduction in the electricity sector as a component of a clean fossil fuel strategy. Estimations come up with an increase of the electric generation cost by about 0.01 to 0.05 US$ per kWh. A significant level of CCS is expected provided CO_2 prices reach 25 to 30 US$ per ton of CO_2. Real costs will also depend on the legal or regulatory framework which is just developed on an international scale, e.g. in EU (EC, 2008).

CCS technologies were first proposed in the 1990's as an atmospherically viable option. Currently several research and demonstration plants are in operation. *ELSAM (CASTOR)* started the operation of a demonstration power plant with a tail end capture for about 2.5 MW thermal load. *Vattenfall (ENCAP)* decided on building a demonstration plant for integrated capture in the case of a coal fired oxyfuel-boiler technology with about 30 MW (Span, 2006). The European Commission decided to support the construction of up to 12 demonstration facilities. As an example, in a 60,000 t CO_2 storage research project named *CO2SINK* in Ketzin (Germany) long time storage in a 800 m deep sandstone rock is studied by geo-physical and geo-chemical monitoring systems (BMWi, 2007).

Several industrial-scale storage projects are in operation, which are an offshore saline formation in Norway, an oil recovery project in Canada, and an operation in a gas field in Algeria. For power production, *RWE* (RWE, 2006) gave notice of commercial operation of a 450 MW plant with CO_2 capture for 2014. The economics of the plant will depend solely on the price of emission certificates. More than EUR 20 per ton CO_2 avoided is required. *BP* will start commercial operation of a natural gas fired 350 MW power plant in 2009. The project mainly aims at CO_2 production for enhanced oil recovery, power is a "by-product". Investment is about EUR 500 Mio. A return of about 40 Mio barrels of oil is planned (Span, 2006).

10 Climate Impacts and Emission Mitigation of Industrial Production

10.1 Relevance and Trends of Industrial Sector Emissions

World's industrial sector emissions of GHGs, which comprise CO_2 from energy and from non-energy use of fossil fuels and renewable fuel sources, as well as non-CO_2 GHG emissions, amounts 7,200 Mio t CO_2-eq. and 12,000 Mio for direct and for total emissions including indirect emissions, respectively. A specification after sources is given in table 10.1.

Table 10.1 Industrial GHG sources and amounts (world's totals) (IPCC, 2007c)

GHG source	Amount Mio t CO_2-eq. (reference year)	Specifications
Energy-related CO_2	9,900 (2004)	In 1971 the value was 6,000 Mio t CO_2-eq., which is equivalent to a 40 percent growth.
		Industrial sector's share of global energy use was 37 percent in 2004
		Share of developed nations, transition economies and developing nations was 35, 11 and 53 percent, respectively.
CO_2 from non-energy use of fossil and non-fossil fuels	1,700 (2000)	
Non-CO_2	430 (2004)	Includes CH_4, N_2O, PFCs and HFCs

Not more than 4 percent of world's GHG emissions are directly process related. In the EU industrial production contributes by about 10; in the U.S. by about 5 percent. This is a relatively small position compared to the energy related processes (see figure 9.1). Nevertheless, GHG emission reduction is important also for the industry.

Emissions considered in industry are released as by-products of the reactions which take place in the industrial processes, when raw material is transformed from one state to another in the end product. In addition to that, the industrial processes are also characterized by energy consuming process steps such as heating or cooling for best process conditions, or electrical energy for smelting processes (as in the case of aluminum production), or stirrer power for stirring of the process fluids (as in the case of bioprocesses, liquid-liquid reactions or solvents), and others. Such energy related emissions are considered independent of the process and are not part of the process emission balance.

On the other hand there is a large difference in the GHG potentials of different technologies. Reduction measures are especially important for the industries with largest emissions. A survey of the situation in the USA where industrial processes in 2005 emitted more than 330 Mio t of CO_2-eq. is given in table 10.2.

Table 10.2 Total emissions from selected industrial processes in the U.S. industry, 2005, (Mio t CO_2-eq.) (EPA, 2007b)

Process	CO_2	CH_4	N_2O	HFCs, PFCs, SF_6	All
Substitution of Ozone Depleting Substances				123.3	123.3
Iron and steel production	45.2	1.0			46.2
Cement manufacture	45.9				45.9
Ammonia manufacture and urea application	16.3				16.3
Nitric acid production			15.7		15.7
HCFC-22 production				16.5	16.5
Electrical transmission and distribution				13.2	13.2
Lime manufacture	13.7				13.7
Aluminum production	4.2			3.0	7.2
Limestone and dolomite use	7.4				7.4
Adipic acid production			6.0		6.0
Semiconductor manufacture				4.3	4.3
Petrochemical production	2.9	1.1			4.0
Soda ash manufacture and consumption	4.2				4.2
Magnesium production and processing				2.7	2.7
Titanium dioxide production	1.9				1.9

Among the traditional industrial branches iron and steel production as well as cement manufacture are main emittents. 50 percent reduction happened between 1990 and 2005. A much larger part is emitted by the so called new processes for the replacement of ozone depleting substances (ODSs) by HFCs and PFCs. By 2020, ODS substitutes are expected to account for 60 percent of all industrial emissions. Substantial increases of GHG emissions are projected from HCFC production and electric power systems with growth rates of about 60 and 80 percent, respectively. Semiconductor manufacturing will show almost doubled emissions, despite the adoption of mitigation measures (WWF, 2006). Current emissions and mitigation potentials will be considered in chapter 10.2.

In the majority of all industrial processes CO_2 is emitted. Some typical GHGs other than CO_2 emitted from selected industrial processes which were displayed already in table 9.2 are given in detail in table 10.3.

Table 10.3   **Industrial processes and typical non-CO_2-emissions (EPA, 2006)**

Product	Emitted GHGs
Adipic and nitric acid	N_2O
Substitutes for ozone depleting substances	HFCs, PFCs
HCFC-22	HFCs
Electric power systems	SF_6
Primary aluminum	PFCs
Semiconductor	HFCs, PFCs, SF_6
Magnesium	SF_6
Other miscellaneous industrial products	CH_4, N_2O

Table 10.4 displays specific emissions of CH_4 and N_2O and the resulting total specific greenhouse gas potentials for selected industrial processes.

Table 10.4   **N_2O and CH_4 emissions (Schön, 1993) and GHG potentials of selected processes**

Product	Emissions (kg per ton of product)		Total GWP (t CO_2-eq./t)
	N_2O	CH_4	
Ammonia		10.1	0.25
Nitric acid	3.1-6.2		0.92-1.85

Product	Emissions (kg per ton of product)		Total GWP (t CO_2-eq./t)
	N_2O	CH_4	
Adipic acid	333		99
Methanol		9.1	0.23
Oxo-synthesis products		3	0.07
Acetic acid		9	0.23

Gases other than CO_2, CH_4 and N_2O influence the overall climate effects of industrial processes as well, especially the anthropogenic (man-made) chlorinated and fluorinated hydrocarbons, namely hydrofluorocarbons (HFCs) and perfluorocarbons (PFCs), but also sulphur hexafluoride (SF_6). They are summarized under the term High Global Warming Potential Gases (high GWPs). They are used as substitutes for a group of so called ozone depleting substances (ODSs) which have been phased-out in industrialized countries under the *Montreal Protocol on Substances that Deplete the Ozone Layer.* This usage is growing rapidly. From a current value of nearly 300 Mio t CO_2-eq. a rise to about 700 Mio t is envisaged by 2020. But only the substance group is to be characterized as an interim substitute in many applications, for they will also be phased-out under the provisions of the Copenhagen Amendments to the *Montreal Protocol.*

In some ODSs replacement applications, such as solvent cleaning or aerosol applications, the substitutes are emitted immediately, whereas in others the substitutes are replacing ODSs' in equipment such as refrigerators or air conditioning applications, and are only slowly released.

Moreover, they are employed and emitted by important industrial processes, such as aluminum and HCFC-22 production, semiconductor manufacture, and magnesium metal production and processing. They are also generated through electric power transmission and distribution facilities (see table 10.5).

Table 10.5   High GWPs and their current industrial use (EPA, 2006)

Chemical	Use
Hydrofluorocarbons (HFCs)	
Several HFCs	Foam blowing agent and refrigerant, fire suppressant, propellant in metered dosed inhalers and aerosols, plasma etching and semiconductor production

Chemical	Use
Perfluorocarbons (PFCs)	
CF_4, C_2F_6	Byproduct of aluminum production, plasma etching and cleaning in semiconductor production and low temperature refrigerants
C_3F_8	Low-temperature refrigerant and fire suppressant. Used in plasma cleaning in semiconductor production
Sulfur Hexafluoride (SF_6)	
SF_6	Cover gas in magnesium production and casting, dielectric gas and insulator in electric power equipment, fire suppression discharge agent in military systems, atmospheric and subterranean tracer gas, sound insulation, process flow rate measurement, medical applications, and formerly an aerosol propellant. Used for plasma etching in semiconductor production
Hydrofluoroethers (HEFs)	
$C_4F_9OCH_3$	Cleaning solvent and heat transfer fluid

In addition to such greenhouse gases which directly influence climate factors, many industrial processes generate indirect greenhouse gases which occur when chemical transformations involving the chemical substance produce greenhouse gases. Another indirect effect occurs when the gas considered influences other climate relevant processes such as lifetime of atmospheric GHGs.

The most important indirect industrial greenhouse gases are NO_x, carbon monoxide (CO) and non-methane volatile organic carbon compounds (NMVOCs). They are produced in chemical and allied product manufacturing, metal processing, during storage and transport, as well as by health services, cooling tower operation, fugitive dusts, various incomplete combustion processes, and accidental or catastrophic releases.

10.2 Conseqences of Climate Change for Industry

By the emissions discussed industry actively influences the climate change. On the other hand, industry itself is influenced passively by climate effects. One main factor could be the situation of infrastructure which may be influenced or even destroyed by weather events such as snow, floods, or low water levels in rivers, which make shipping of goods untenable. Low water supply or higher water temperatures make process cooling and environmental activities more difficult.

Moreover, industrial activities would be affected through the impact of government policies pertaining to climate change, such as carbon taxes which increase the material and energy costs. They could also be affected through a changed consumer behavior. An example is clothing, the choice of which depends on the temperature, and more warm-weather clothing might be ordered in cold climates, and vica versa. Climatic impacts on natural resources may influence manufacturing that depends on that type of resource, e.g. food and renewable material processing, which depend on the agricultural yields which may change through climate factors.

This aspect of climate influence on industrial processes is not yet fully clarified, however considered of growing concern.

10.3 Climate Impacts and Emission Mitigation of Selected Industrial Processes

10.3.1 Production of Iron and Steel

World raw steel production was 1,200 Mio t in 2006 (USGS, 2007). Markets are growing very fast, especially in China, where annual production has reached about 420 Mio t in 2006. On a global scale steel production GHG emissions are estimated to be 1,500 to 1,600 Mio t CO_2-eq., including emissions from coke manufacture and indirect emissions due to power consumption. This is equal to 6 to 7 percent of world's anthropogenic GHG emissions. Chinese steel production accounts for more than 10 percent of countrywide GHG missions.

Iron and steel production is climate relevant for two reasons: First, it is an energy intensive process and thus uses large amounts of fossil energy sources, which result in CO_2 emissions. In addition, during the different process steps, non energetic production related emissions of direct climate gases such as CO_2 and CH_4 occur. Moreover other pollutants with direct greenhouse gas effects are emitted (see table 10.6).

Table 10.6   **Emission of selected pollutants in steel production (BREF, 2002)**

Pollutant	Amount (g/t liquid steel)
PAH	200
VOC	90

Pollutant	Amount (g/t liquid steel)
CO	14,900
NO_x	1,000
SO_2	930

The following details of steel production elucidate the sources of emissions:

- Iron is produced by reducing iron oxide from iron ore using metallurgical coke in a blast furnace. Iron is introduced in the form of raw iron ore, pellets, briquettes, or sinter material. The result of this first production step is an impure pig iron. It is the raw material for steel production by specialised steel making furnaces, and for the production of iron products in foundries. During pig iron production CO_2 and CH_4 are emitted.
- Metallurgical coke which is needed as a reducing agent is produced by heating coking coal in a coke oven in a low oxygen environment. This process is a non-energetic one. Thus, its emissions are considered in the greenhouse balance of steel production. During the process, volatile compounds of the coking coal are driven off as a coke oven gas. When applied in the blast furnace, the metallurgical coke is oxidized. The result is reduced iron and CO_2.
- When volatile compounds resulting from metallurgical coke production condensate tar products are generated. Coal tar is the raw material for the production of anodes for electrolytic processes, such as primary aluminum production, and for several other coal tar products. During the processes, CO_2 and CH_4 are emitted. The relation between these two is about 50:1.
- The majority of the CO_2 emissions in the iron and steel production come from the production of pig iron (using metallurgical coke). Smaller amounts originate from the removal of carbon during the steel production. Carbon is also stored in the products, i.e. iron (about 4 percent carbon) and steel (about 0.4 percent carbon).
- Methane which is produced during the processes for coal coke, sinter, and pig iron, is mostly emitted via leaks in the production, only partly through the emission stacks of the plants. Thus a treatment of methane is difficult. The emission factors are 0.5 g CH_4/kg produced coal coke, 0.9 for pig iron, and 0.5 for sinter.

Resulting total GHG emissions in steel production are characterised by a high variability depending on the production technology applied. The differences are based on the production route used, fuel mix and production energy efficiency, carbon intensity of the fuel mix, as well as electricity carbon intensity.

Compared to an average emission value of world steel production EU and USA both reach only 50, and Japan only 25 percent, however Russia and China up to 350 percent due to out-date-technology. This fact indicates a high improvement potential by modernisation of plants. In Germany, long term improvement activities of steel making technology resulted in a reduction from 2.5 t CO_2-eq. per t of pig iron in 1960 to a value of 1.34 in 2007 (Amelung, 2007). After technology assessment studies a value between 0.5 and 1.5 seems possible. However economic constraints will lower this potential.

Technological improvements refer to reduction of carbon dioxide emissions by CCS (see chapter 9.4), energy efficiency improvements, as well as fuel switching. Energy related measures include enhancing continuous production to reduce heat loss, recovery of waste energy and process gas, or preheating in case of the use of scrap metal. Switching to alternative fuels such as oil and gas could reduce specific CO_2 emissions. Other recent options are the use of pre-treated waste components, such as recycled plastics or refuse derived fuel (RDF) (see chapter 12.6). Moreover renewable charcoal is a traditional alternative to coke. By use of hydrogen iron ore could be reduced with much lower CO_2 emissions.

Long term total mitigation potential is estimated in the range of 0.32 and 0.76 t CO_2-eq./t steel.

10.3.2 Cement and lime manufacture

10.3.2.1 Cement manufacture

Cement is a finely ground grey powder of inorganic non-metallic nature. After mixing with water it forms a paste which sets and hardens due to the formation of silicate hydrates from cement constituents. Cement is a critical element of the construction industry. World cement production in 2006 was about 2,500 Mio t (USGS, 2007). It is produced in nearly 40 countries all over the world. The production is growing heavily due to the global economic development, e.g. in China in 2006 a capacity of 1.1 billion t was achieved (USGS, 2007). No shortage of the raw material is envisaged, for limestone is abundant all over the world.

The cement industry strongly contributes to the global CO_2 emissions with an amount of 5 percent. In the USA, cement is one of the largest sources of industrial CO_2 emissions (see table 10.2). At the current cement production of 94 Mio t of portland cement and 6 Mio t of masonry cement a total of 45.9 Mio t CO_2-eq. is emitted.

Comparable with steel production, climate relevant emissions originate from direct energy consumed in making cement, as well as from the chemical processes during the reaction. An amount of 60 to 130 kg fuel oil and 110 kWh of electrical energy per ton of cement are used. 50 percent of the greenhouse gas emissions are due to the chemical process, and 40 percent to burning fuels. The remainder splits into transport and electricity needs.

Specific production processes are as follows: In the first process the raw material which is limestone (calcium carbonate $CaCO_3$) is heated at a temperature of about 1,300°C in a cement kiln ("calcination"). This results in lime (calcium oxide CaO) and CO_2. The amount of CO_2 released is directly proportional to the lime-stone input. In the next step, as an intermediate product, the so-called clinker is produced through the combination of lime with silica-containing material. An average of 0.525 t CO_2 is emitted per ton of clinker produced. The clinker is cooled and then mixed with small amounts of gypsum, which results in portland cement. Masonry cement for construction needs is produced by the addition of more lime. This results in additional CO_2 emissions.

Also methane is emitted, but in very small amounts, which are in the range of 0.01 percent of the CO_2 emissions.

The reduction of CO_2 emissions is a first priority in the cement industry. In a *Cement Sustainability Initiative* a standard ruling for the cement production process and measures to reduce emissions were prepared (CSI, 2005).

One important measure to reduce the climate gas emissions is the substitution of traditional fossil fuels by industrial wastes, which is also one of the most efficient practices of disposing waste components. On the other hand waste components not only substitute energy but also raw material so that a double effect is achieved as long as quality requirements of the target product are met.

A survey of secondary fuels which include such components as used plastics insulation, shredded plastics, paper fractions, and municipal solid waste as well as refuse derived ("secondary") fuels (RDF; see chapter 12.6) is given in table 10.7.

Table 10.7   Secondary fuel use in cement production (UBA, 2007)

Secondary fuel source	CO_2 emission factor (kg/ TJ)	Biogenic mass fraction (percent)
Recycled tyres	97.32	27
Recycled oil	78.69	0
Commercial waste – paper	64.88	91
Commercial waste – plastic	83.07	0
Commercial waste – packaging	56.85	40

Secondary fuel source	CO_2 emission factor (kg/ TJ)	Biogenic mass fraction (percent)
Textile waste	63.29	70
Commercial waste – other	68.13	52
Animal meals and fats	74.87	100
Processed municipal waste	59.85	55
Waste wood (wood scraps)	95.06	100
Solvents (waste)	71.13	0
Carpet waste	80.42	36.5
Bleaching clay	82.26	0
Sewage sludge	95.11	100
Oil sludge	84.02	0

German cement industry annually applies about 2.8 Mio t of such secondary fuels. Similar activities can be observed in other countries such as Japan and India for the use of waste material, agricultural wastes, sewage sludge and a wide range of organic liquids and solvents. On the level of single companies or cement plants the usage of waste components in an amount of more than three quarters of the fuel was reported (ICCP, 2007c).

In some cases, extra climate benefits are achieved from reduced CO_2 emission per energy unit which is due to the biogenic organic carbon content (see table 10.7) which is considered climate neutral (see chapter 12). In the case of refuse derived fuel (RDF) from municipal solid waste (MSW) about 60 percent is of biogenic origin.

As another climate control option considerable amounts of foundry sand, which is a by-product of steel making, can be used in cement production in place of cement clinkers. One ton of cement clinker can be replaced by one ton of foundry sand, and this relationship defines the pertinent CO_2 emission reduction. In Germany, in 2004, 5.11 Mio t of cement clinker were replaced by foundry sand. This is also a waste reducing activity (UBA, 2007).

The total long term mitigation effect is estimated between 0.65 and 0.89 t CO_2-eq./t cement (IPPC, 2007c).

10.3.2.2   Lime manufacture

The term lime refers to a broad variety of chemical substances, which includes high-calcium quicklime (calcium oxide, CaO), hydrated lime (calcium hydroxide, $Ca(OH)_2$, dolomite quicklime ($CaO*MgO$), and dolomite hydrate (e.g. $Ca(OH)_2*Mg(OH)_2$).

Lime is not only used in the cement production, but is also a manufactured product, which has many industrial, chemical, and environmental applications, mainly in steel making, as a purifier in metallurgical furnaces, in cleaning (desulfurisation) of flue gas (FGD) from coal-fired electric power plants, in construction, in water purification, or as raw material in glass manufacturing and magnesium production from dolomite.

World annual lime production amounted to about 130 Mio t in 2006 (USGS, 2007), with China and USA as main producers, which produced 25 and 21.2 Mio t, respectively, followed by Russia (8.5), Japan (8.9), and Germany (6.8 Mio t). Lime ranks high among the most important chemicals, in the U.S. industry it was historically fifth in total production of all chemicals.

The main technological step – in analogy to the cement production – is the calcination, where CaO is produced and CO_2 is emitted to the atmosphere. Theoretical process emissions are 0.785 t CO_2 per ton of calcium oxide and 1.092 t CO_2/t magnesium oxide produced. In efficient lime kilns about 60 percent of the emissions are due to the chemical processes. In Europe they are estimated at 0.750 t CO_2/t lime. The value can be up to 2-3fold, e.g. in small vertical kilns in Thailand. Emissions from fuel depend on the kiln type, energy efficiency and fuel mix and are 0.2 to 0.45 t CO_2/t lime (IPPC, 2007).

More efficient and better managed kilns are a pre-condition of emission reductions. Further reductions are possible by switching to low-fossil carbon fuels as in case of cement. For small scale facilities the use of solar energy seems promising.

A major reduction is possible by recovering of CO_2 for use in sugar refining and for the production of precipitated calcium carbonate (PCC), which is applied as special filler in premium-quality coated and uncoated papers. In the U.S. industry in 2004 1.125 Mio t CO_2 were used in this way which is equal to about 7 percent of the total U.S. CO_2 emission in the lime industry. 90 percent of the CO_2 input in the U.S. sugar refining and PCC production were from this source.

10.3.3 Ammonia manufacture and urea application

Ammonia and urea are nitrogen fertilizers for application in agriculture to improve plant yields. Annual world ammonia production in 2006 was about 122 Mio t (USGS, 2007).

At present, feed-stocks of ammonia production are natural gas, but also petroleum coke. In the case of production from natural gas, there are five main process steps, including a primary and a secondary reforming and a shift reforming process, by which CO_2 is removed from the process. During the following ammonia synthesis, NH_3 is formed by a catalytic process from H_2 and N_2.

CO_2, together with process impurities, is a constituent of the waste gas. It is washed out by a scrubber from which it is released into the atmosphere during regeneration of the scrubber solution. A part of the CO_2 is used as a raw material in the production of urea ($CO(NH_2)_2$) together with ammonia. The carbon in the urea is released into the environment after application of the urea fertilizer in agriculture. Hence, the whole amount of CO_2 produced in the ammonia synthesis is finally emitted into the atmosphere. For greenhouse gas budgets these CO_2 emissions are allocated to ammonia or urea production according to the amount of both fertilizers.

The emission factor is 1.2 t CO_2-eq. per t of NH_3 in the case of natural gas feedstock. For each ton of urea 0.73 t CO_2-eq. are emitted. The long term mitigation effect is about 0.5 CO_2-eq./t (IPPC, 2007c).

10.3.4 Aluminum production

Aluminum is a light weight metal with high corrosion resistance and high heat and electric conductivity. Its annual global production in 2005 was about 31.2 Mio t (USGS, 2007) with highest capacities in China (7.2 Mio t), Russia (3.7), Canada (2.8), and the USA (2.5). Production is expected to grow at a rate of 3 percent per year during the next decade.

In quantity and value aluminum is the second in the range of metals after steel and is important in all segments of the economy. Uses include transportation (aircraft, automobiles, bicycles), construction (windows, doors), packaging (cans, foil), and consumer durables for daily life, such as kitchen ware and material. By the so-called eloxation a thin surface cover is generated which improves the primarily high corrosion resistance and makes aluminum even better applicable in the construction sector.

In nature, aluminum occurs as a low soluble oxide and as a silicate. The production of primary aluminum from the ores (especially bauxite) is very energy consuming (via electrolytical processes), so that a reduction of the power input seems to be the best choice of process and climate improvements. However latest worldwide efforts to reduce electricity needs in primary aluminum production were without result due to the fact that the production is currently optimised, and the energy need is close to the theoretical minimum (by a factor of 2) (Schön, 2004).

The best option to reduce power input for total aluminum production is recycling, since the energy need for recycled aluminum is only 8.5 percent of the primary production for Western European or German energy mix, respectively. Sources for recycled aluminum include automobiles, windows and doors. However, recycling of aluminum cans has the highest profile. In the USA in 2006 aluminum recovered from purchased scrap was about 3 Mio t, of which 64 percent came from manufacturing scrap and 36 from discarded aluminum products (old scrap). Aluminum from old scrap was equivalent to 16 percent of apparent consumption (USGS, 2007). For details of aluminum recycling effects on CO_2 emissions see chapter 12.

In additions to the energy aspect also process related emissions of CO_2 and PFCs occur, especially perfluormethane (CF_4) and perfluoroethane (C_2F_6), both characterised by very high global warming potentials (see table 7.1).

The emission of CO_2 occurs during the aluminum smelting process, when aluminum oxide from the ores is reduced (Hall-Heroult reduction process) through electrolysis in reduction cells. The cells contain a molten bath of cryolite (Na_3AlF_6) which is of natural or synthetic origin. For the cathode in the electrolytic process a carbon lining is used. As anode, also carbon containing material is applied. Carbon is oxidised during the process and emitted into the atmosphere as CO_2.

The amount of CO_2 released is approximately 1.5 t/t aluminum produced. In another technology (so-called Soderberg cell) 1.8 t/t are released. A technology shift from this cell would reduce the emissions by 20 percent.

Aluminum production industry in addition to CO_2 is a source of PFC emissions. The reason is the so-called anode effect by which the voltage in the electrolysis bath rapidly increases due to reduced levels of the smelting bath. Then reactions of carbon and fluorine of the molten cryolite bath take place, and fugitive emissions of CF_4 and C_2F_6 occur. Their magnitude depends on the process conditions and can be massively reduced if anode effects are minimised by better control technologies. In the U.S. aluminum industry PFC emissions declined by a factor of 6.6 in the last 15 years. The relation of CO_2 emissions to PFC emissions currently is about 1:0.7 compared to 1:2.6 in 1990. Further drastic reduction is expected by use of an inert anode type that would eliminate CO_2 and PFC emissions from the smelting process. Commercially viable results are expected by 2020.

Current climate efficient measures mostly rely on enhanced recycling of scrap aluminum.

10.3.5 Carbon Dioxide Use

CO_2 is used for food processing, in chemical production, for beverages, refrigeration, as a greenhouse fertilizer, or in the petroleum industry for enhanced oil recovery (EOR). In the case of EOR, CO_2 is injected into the underground to rise the reservoir pressure, so that additional oil can be extracted (see also chapter 9.4).

CO_2 is a by-product of many industrial processes (e.g. ammonia production, fossil fuel combustion, bioethanol production, lime processing), but also emanates during extraction of crude oil and natural gas of which it is a naturally occurring constituent. Other feed-stocks of CO_2 are natural CO_2 reservoirs.

The methodology for the accounting of CO_2 used in industry is not yet fully available. Typically the following assumptions are made: In the case of enhanced oil recovery the CO_2 applied is assumed to remain sequestered in the reservoirs (see chapter 9.3; CCS). For all other CO_2 uses it is assumed to be released into the atmosphere during or after the process.

Fossil fuel burning related CO_2 is not considered in this chapter.

Under these conditions 1.2 Mio t CO_2-eq. are emitted in the USA. This is less than one percent of the total greenhouse gas emissions there.

10.3.6 Semiconductor manufacture

Semiconductors are produced using different long-lived fluorinated gases during plasma etching (patterning) and plasma enhanced chemical vapour deposition (PECVD). About one hundred process steps requiring fluorinated gases are used to produce the semiconductor products such as devices or chips from silicon wafers.

Plasma etching is applied to provide pathways for conducting substances which connect the circuit components of the semiconductors. Plasma-generated fluorine atoms are used. These atoms chemically react with exposed dielectric films. During this process certain portions of the film are selectively removed. Some residual undissociated fluorinated gases remain. They together with removed material are partly emitted as waste gas, and are partly treated in emission abatement systems.

PECVD chambers are periodically cleaned by using fluorinated gases, which in plasma, are converted to fluorine atoms. They move away the residual material from chamber walls, electrodes and hardware. Residues and reaction products, e.g. CF_4, are emitted.

For these purposes, depending of the specifity of the products, mainly the following gases are used: Trifluoromethane (HFC-23, CHF_3), perfluoromethane (CF_4), perfluoroethane (C_2F_6), nitrogen trifluoride (NF_3), and sulphur hexafluoride (SF_6), moreover perfluoropropane (C_3F_8), and perfluorocyclobutane (c-C_4F_8).

Climate Impacts and Emission Mitigation of Selected Industrial Processes   121

About 500 t of these substances were applied in the U.S. semiconductor industry in 2004. This seems to be a relative low amount. But these gases are highly potent greenhouse gases (see table 7.1), and thus end up as about 4.7 Mio ton CO_2-eq. The numbers increased during the last 15 years, due to the growth of the industry and the higher complexity of the semiconductors, which needs more PFCs. But there has been a declining growth rate of PFCs in recent years due to process optimisation and abatement technologies. The decline in growth was about one-third during the last 5 years.

10.3.7 Nitric and adipic acid production

Both nitric and adipic acid are feedstocks or components in a variety of commercial products' manufacture. The processes are the major emittents in industry of N_2O.

10.3.7.1   Nitric acid

Nitric acid (HNO_3) is an anorganic compound. It is used primarily to make synthetic commercial fertilizer, but is also a major component in the production of adipic acid and explosives. The usage as a fertilizer will arise worldwide but will decline in Western Europe, due to the concerns about nitrates in groundwater supply.

Nitric acid is produced by the catalytic oxidation of ammonia, where N_2O is formed as a by-product and released from reactor vents into the atmosphere. Currently there is no control measures aimed at eliminating N_2O from the process, whereas the waste gas stream may be cleaned of other pollutants such as nitrogen dioxide (NO_2).

10.3.7.2   Adipic acid

Adipic acid is a white, crystalline solid. Chemically it is also called hexanedioic acid (hexane-1,6-dioxic acid), which is a C6 straight-chain dicarboxylic acid. It is only slightly soluble in water but very soluble in alcohol and acetone.

The annual world production was about 3.8 Mio t in 2003 (EPA, 2005b). One-third is produced in the USA. It accounts for 90 percent of commercial nylon production which is further processed into fibers for applications in carpeting, automobile tire cord, and clothing. Furthermore adipic acid is used for plasticizers and lubricants components, and for making polyester polyols for polyurethane systems. Food grade adipic acid is used as a gelling aid, acidulant, leavening and buffering agent. Its derivates are used in making flavoring agents, internal plasticizers, pesticides, dyes, textile treatment agents, fungicides, and pharmaceuticals.

Commercial adipic acid is mostly produced from cyclohexane through a two-stage oxidation process. The first involves the catalytic reaction of cyclohexane with oxygen to produce a mixture of cyclohexanol and cyclohexanone. Afterwards, adipic acid is formed by another catalytic reaction of the mixture with nitric acid and air.

Due to the use of nitric acid in this process, N_2O is generated at a rate of 0.333 t per ton of adipic acid. It is emitted by the waste gas stream. In modern plants the waste gas is treated by an abatement system which destroys pollutants by catalytic or by thermal reactions, with an efficiency of 95 and 98 percent, respectively. Due to the installation of these technologies in most plants, the N_2O emissions from adipic acid production have been reduced by two-thirds since 1990. In 2000 the total emissions by adipic acid production were about 50 Mio t of N_2O. Despite positive effects in reducing specific emissions by abatement technologies, a rise of the emissions by about one-third to about 67 Mio is envisaged due to rising worldwide production capacities. The current total emissions caused by adipic acid in the U.S. industry is about 5.7 Mio ton CO_2-eq.

11 Climate Effects of Agricultural Processes

11.1 Overview on Agriculture and Climate Interaction

The agricultural production system is strongly influenced by several climate factors. Therefore it passively suffers or benefits from climate change. On the other hand it actively influences climate change due to two separate aspects: During most of its production steps greenhouse gases are emitted, which adversely influences the climate. However, agriculture is the producer of renewable crops which replace fossil fuels and supply industry with renewable raw material. This contributes to heavily reduced climate impacts mostly of energy and traffic sectors. To define the net effect of agriculture on climate factors, a complex analysis is necessary.

The following figures illustrate the agricultural emissions of GHGs on a global scale: Agricultural activities accounted for 10-12 percent of total anthropogenic GHG emission in 2005, which is about 5,100 to 6,100 Mio t CO_2-eq./a. The number indicates higher emissions compared to industry. In the case of the USA where nearly 540 Mio t CO_2-eq. are emitted, the value is 7.4 percent of the total national greenhouse gas emissions compared to only 4.6 percent from industry (without energy sector) (2005; EPA, 2007b).

Primary greenhouse gases from agriculture are methane (CH_4) and nitrous oxide (N_2O). CH_4 contributes 3,300 Mio t CO_2-eq./a and N_2O about 2,800 Mio t CO_2-eq./a, which is equivalent to about 50 and 60 percent of global anthropogenic emissions by these substances, respectively.

Methane is emitted from enteric fermentation of domestic animals, especially beef and dairy cattle, which are the largest emitters of CH_4 due to their ruminant digestion system. Rice cultivation is 11 percent on global average. It dominates in the group of developing countries where it accounts for more than 90 percent of the world total. Rice production emissions are of minor importance in developed countries due to lower production numbers as well as different cultivation methods. Also burning of biomass (12 percent) is mostly due to such activities in the developing rather than in the developed countries. Another 7 percent originate from the management of manure from livestock.

N_2O is predominantly released by agricultural soil management activities. In the case of the USA it accounts for more than five percent of total national emissions or three-quarters of the total national N_2O emissions. Other sources of N_2O are manure management in animal breeding and agricultural residue burning. It is to be mentioned that in the case of residual burning CO_2 emissions are not counted as climatically relevant, since the assumption is made that carbon released from biomass burning into the environment as CO_2 will be reabsorbed in the following seasons. Only methane, N_2O, CO and NO_x, are considered in the budget of biomass burning.

How much its five main GHG sources contribute to the total of modern agriculture emissions through CH_4 and N_2O in the global mean as well as in a highly developed agriculture (in the case of the USA) is given in table 11.1 in relative numbers. For the USA the percentage is based on a national total of about 540 Mio t CO_2-eq. in 2005 (EPA, 2007b).

Table 11.1   GHG emissions in agriculture (Global and U.S. agriculture, 2005; percent)

GHG source	U.S. agriculture (EPA, 2007b)		World (IPPC, 2007c)
	CH_4	N_2O	N_2O and CH_4
Enteric fermentation of livestock	20.9	0	32
Manure management	7.7	1.8	7
Rice cultivation	1.3	0	11
Agricultural residues burning	0.2	0.1	12
Agricultural soil management	0	68.1	38

The recent figures in non-CO_2 GHG emissions are a consequence of a strong increase observed during the last two decades when the average emission increased by about 60 Mio t CO_2-eq. annually (from 1990 to 2005). Nearly 90 percent were due to biomass burning, enteric fermentation and soil nitrogen emissions. A decrease was seen in the developed countries. Of future non-CO_2 emissions estimates are about 8,300 Mio t CO_2-eq. by 2030 (IPCC, 2007c).

As was mentioned, emissions by agriculture are the one side of the medal by which it has high climate impacts. On the other hand agriculture is highly sensitive to climate variability and weather extremes such as severe droughts, floods and storms, which are critical to farm productivity. Climate variability and change also modify the risks of fires, pest and pathogen outbreak, negatively affecting food, fiber and forestry (IPCC, 2007c). Table 11.2 connects climate change and agricultural productivity (EPA, 2007c).

Table 11.2   Effect of climate change factors on agriculture

Climate factor	Effects	Influence
Average temperature increase	Lengthen the growing season in regions with a relatively cool spring and fall	+
	Adversely affect crops in regions where summer heat already limits production	-
	Increase soil evaporation rates	-
	Increase the chances of severe droughts	-
Change in rainfall amount and patterns	Changes in rainfall affects soil erosion rates and soil moisture	-
	Precipitation will increase in high latitudes	+
	Precipitation will decrease in most subtropical land regions	-
	Number of extreme precipitations will increase	-
Rising atmospheric concentrations of CO_2	Increasing atmospheric CO_2 levels can act as a fertilizer and enhance the growth of some crops such as wheat, rice and soybeans	+
Pollution levels such as tropospheric ozone	Higher levels of ground level ozone limit the growth of crops	-
Change in climatic variability and extreme events	Changes in the frequency and severity of heat waves, drought, floods and hurricanes, are foreseen by global climate models	-

Factors given in table 11.2 must not be considered separately. Positive effects by one factor may be offset by others. While food production may benefit from a warmer climate or an elevated CO_2 level, high levels in ozone concentration in atmosphere and lower precipitation may reduce yields. Moreover, regional effects must be considered which cannot be forecasted by global models. In general, after recent studies it is expected that agriculture in industrialised regions will be less vulnerable to climate change than in developing countries.

In North American rain fed agriculture the climate change will likely increase yields by 5 to 20 percent over the next decades, with high spatial variability. Special problems are foreseen in the tropics where agriculture may have little ability to adapt. In certain regions the changes in climate, water supply and soil moisture could make it less feasible to continue crop production. However, agriculture sector's ability to cope with and adapt to climate variability and change will depend not only on changing climate conditions. Adaptation through future changes in technology, management practices, in food and renewable crop demand, in water availability, and soil quality will be crucial (EPA, 2007c).

Though agricultural emissions are not more than 10 to 12 percent of world totals, mitigation in this sector is studied intensively. It concentrates on the i) reduction of emissions, ii) enhancing of removals as well as iii) avoiding or displacing of emissions (IPCC, 2007c). They have to be applied in complex. The total mitigation potential on a global scale was estimated to 4,500 to 6,000 Mio t CO_2-eq. annually if there were no economic constraints. Of this numbers about 90 percent were due to soil carbon sequestration and nine and two percent by methane and nitrous oxide mitigation options, respectively. If constraints by carbon prices would be in the range up to 20 US$, a mitigation potential of about 1,500 Mio to CO_2-eq. in 2030 was calculated by models.

The following chapters deal with the emission situation and the mitigation options for selected agricultural activities.

11.2 Greenhouse Gas Emissions by Livestock

Livestock GHG emissions equate to 1,725 Mio t CO_2-eq. annually on a global scale. This is dominated by methane from ruminants such as cattle and sheep, which accounts for nearly one-third of global anthropogenic emissions of this gas and is the largest methane source globally. In the case of the U.S. livestock total CH_4 emissions in 2005 were about 112 Mio t CO_2-eq. (EPA, 2007b), mostly from beef and dairy cattle (by 95 percent). This is more than the total of the emissions from iron and steel and cement industry. In Germany lifestock methane emissions are close to one (0.87) Mio t CH_4 (UBA, 2007) which is more than 20 Mio t CO_2-eq. and equates to two percent of the total national GHG emissions.

The background of methane production by ruminant livestock is as follows: Nutrients from the food consumed by the animals are decomposed in their digestion systems. During this complex process which partly takes place in an oxygen free atmosphere, anaerobic bacteria produce methane as one of the end-products. As methane producers ruminants like cattle, sheep, goats, and camels dominate. In their rumen which is a type of a fore-stomach, bacteria break down the feed so that it can be absorbed and afterwards metabolized by the following intestinal organs. Ruminants are thus able to digest coarse plant material, such as grass, and other green crops containing high cellulose. Methane produced is exhaled by the animals.

The amount of methane generated primarily depends on the type of the digestive system of the animal. Other factors are amount and composition of the feed consumed. Energy rich feed results in more methane.

Cattle naturally emit between 150 and 250 liters of methane per day (40 to 65 kg per head per year). Under an energy rich feeeding regime, higher values may occur. The CH_4 emission factors of dairy cows in Germany are in the range of 95 kg per head and year with an increase between 1990 and 2005 from 77 to 117.5 (UBA, 2007).

Other kinds of livestock have considerably lower individual emission rates, as given in table 11.3.

Table 11.3   Emission rates of livestock other than cattle (EPA, 2006)

Animal	Emission rate (kg methane per head per year)
Horses	18
Sheep	8 – 10.5
Goats	5 – 6.5
Pigs	1.3
Poultry	0.09

Mitigation proposals imply improved feeding practices and use of specific dietary additives. Better feeding practices include measures such as optimization of protein intake, adding oils or oilseeds, or improving pasture quality especially in low developed agricultures. The use of additives aims at suppressing or avoiding methane generation during the digestion process. For many of the proposed agents a clear benefit was not proven and side effects were observed. In some cases the agents were banned. Vaccination against certain methanogenic bacteria was studied. In the case of sheep it was estimated that a reduction of methane emissions by 20 percent could be achieved. In the case of Australia where the tests were made, livestock GHG emissions by sheep and cattle amount 14 percent of the country's total GHG emissions. A reduction by 0.3 Mio t CO_2-eq. annually was predicted (CSIRO, 2001). However a commercial vaccine is not yet on the market (IPCC, 2007c).

11.3 Climate Effect of Manure Management and Biogas

11.3.1 Manure management

Manure from animal livestock generally contains residual carbon, nitrogen and other substances. During its storage in or outside the stall as well as its application methane and nitrous substances such as N_2O, ammonia, and nitrogen gas are emitted. The extent depends on a range of factors which include animal category, feed composition, but also the manure management type such as liquid or solid handling. Methane is produced from manure in an amount which depends on the concentration of organic residues in the manure. Of nitrogen a portion is excreted into the manure as a consequence of protein digestion and is converted to N_2O by bacterial processes of nitrification and denitrification (for details see chapter 11.4). After measurements by Schön (1993) per cubic meter of cattle manure about 1,300 g methane and 30 gram nitrous oxide were deliberated. Both CH_4 and N_2O contribute to the climate effects of agriculture. In U.S. agriculture, this type of emissions amounts to a total of about 41.3 and 9.5 Mio t CO_2-eq. for CH_4 and N_2O, respectively (EPA, 2007b) or about 10 percent of the total emissions in agriculture.

There is a range of technologies applied to prevent such emissions, including separation of solid components of manure, aeration, anaerobic biogas fermentation of manure or coverage with straw. As is indicated by figure 11.1 the best results are achieved through biogas fermentation where emission is 60 percent lower compared to untreated manure. The coverage with straw results in even higher emissions.

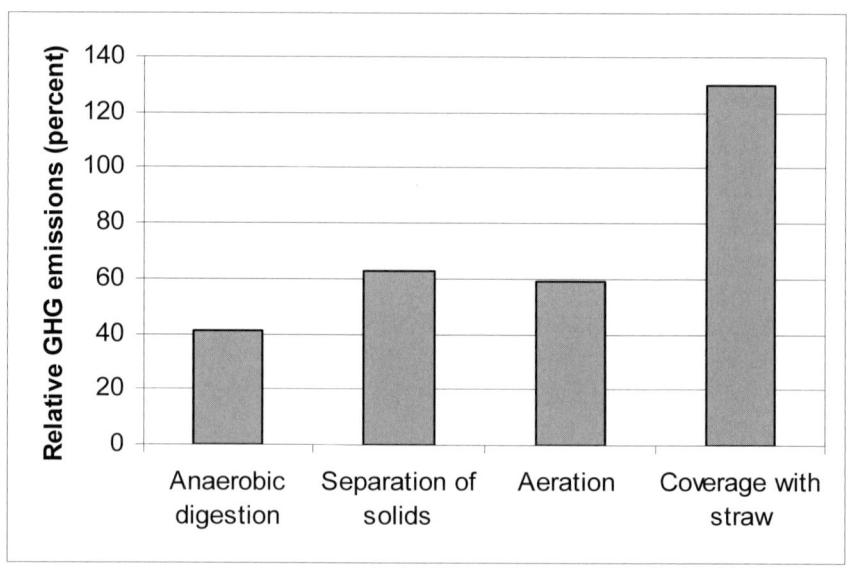

Figure 11.1 **Relative GHG from selected manure treatment technologies**

11.3.2 Biogas production

The microbial processes which naturally take place during metabolizing of manure components can be used in a controlled conversion in a technical process to produce biogas. A combination of positive effects results: Climate related emissions of CH_4 are partly or fully prevented. The quality of the manure and its applicability as a fertilizer are improved. Easily degradable odor producing substances and aggressive organics are metabolized and therefore no longer emitted in large amounts. Thus air pollution is reduced and the fertilizing effect is improved.

It is an even more important benefit that biogas is a renewable energy source which can replace fossil fuels for the generation of electrical power and heat since it causes extra climate benefits by avoidance of fossil fuel burning GHG emissions. Effects are summarized and quantified in table 11.4.

Table 11.4   Climate related effects of biogas production (FVB, 2006)

Consequences of biogas production in agriculture	Quantitative or qualitative effects
De-centralized renewable energy supply	Calorific value (biogas with 60 percent methane): 6 kWh/m^3
	Power potential: 1.8-2 kWh$_{el}$/m^3 biogas
	Mineral oil equivalent: 0.62 l/m^3 biogas
	Performance: 1-1.5 m^3/animal and day
	Biogas from manure of four cattle units equals to the energy need of one household
Reduced climate burdens by climate-neutral energy supply	Installed capacity of 1 kW$_{el}$ avoids 7,000 kg CO_2-eq./a
	Methane emissions by 1,500 kg CO_2-eq. per cattle unit and year are prevented
Use of treated manure as fertilizer	Amount of nitrogen is equivalent to 20 kg N per cattle unit per year
	Odor emissions during application of fertilizer are reduced
Improved sanitary and environmental situation at higher supply security	
Regional economy improved by use of internal resources, empowerment of less developed regions	

The principle technology of biogas production is already in widespread use in developed and developing regions of the world. In rural regions biogas is a traditional de-central energy source for cooking or heating which is often applied in families or households only on the basis of the manure of a low number of cattle applying a relative simple technology. In opposition, the modern biogas strategy aims at producing electricity on the scale of several hundreds kW up to the MW capacity applying a combined heat and power process. A diversity of substrates is used, including manure, organic waste of agricultural or industrial origin, and renewable crops. This type has been established in recent years, e.g. in Germany, where more than 3,700 biogas plants are in operation. The installed electrical capacity is 1,300 MW.

The efficiency of an agricultural biogas process depends on several factors amongst which the type of the raw material and the technological layout are most important. With reference to the raw material the organic content per mass unit is decisive for the biogas yield. Of manure, organic waste and renewable energy crops examples are given in figure 11.2.

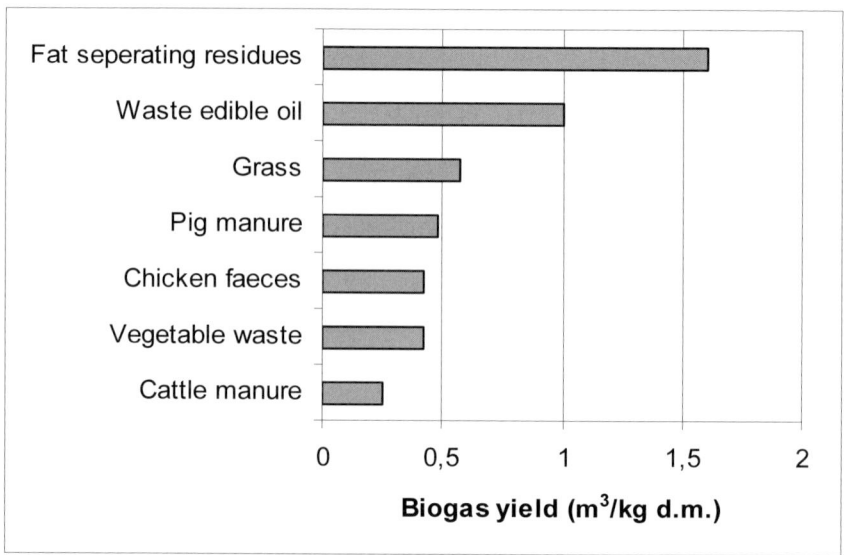

Figure 11.2 Biogas yield of manure and renewable crops

Obviously best results will be achieved if a high proportion of fat residues is fermented. Only cattle manure results in a poor gas yield which is due to the utilization of energy rich substances during rumen digestion already.

A variety of technologies is applied dependent of the type of the substrate and the end usages of generated biogas and processed residues as well as cost and maintenance factors. It necessarily includes equipment for the preparation of the substrate, the conversion in a closed reactor under a mostly oxygen-free atmosphere, as well as the collection and cleaning of the biogas and the conditioning of the residues. The resulting biogas has to be transformed into power in an electrical generator. Only the electricity yield is 30 percent. The larger part will be lost as heat if not a combined heat and power (CHP) process takes place which is typical for modern sustainable solutions.

An example of a layout of such a biogas plant is displayed in figure 11.3. The plant is supplied by manure from 1,500 dairy cattles as well as an annual amount of 1,000 t of renewable crops such as maize. The electrical efficiency is 250 kW in a combined heat and power process. Heat energy is partly utilized for drying processes in a nearby factory. A special reactor design type is used in this facility where the gas is collected and stored under a plastic cover which is blown up by the gas pressure and afterwards transported to the power generator.

Figure 11.3 **Biogas reactor with plastic cover for gas storage (Markert, 2008)**

11.3.3 Case study: Agricultural biogas production in a sports and recreation center

The following case study deals with the establishment of an agricultural biogas plant in a regional sports and recreation centre in South Africa which is for around 5,000 people. The idea behind the establishment of a biogas plant was the intended shift from a fossil fuels based energy supply toward a renewable one operating on the basis of the potentials in respect to agricultural resources, efficiency improvement options, and skills and manpower. Currently the centre is run by electricity from the grid, and coal and gasoline. Electricity is applied for room heating through geysers, lightening, and operation of electrical devices. Coal is for water heating and cooking purposes, and gasoline for transportation. The total actual energy need is summarized in figure 11.4.

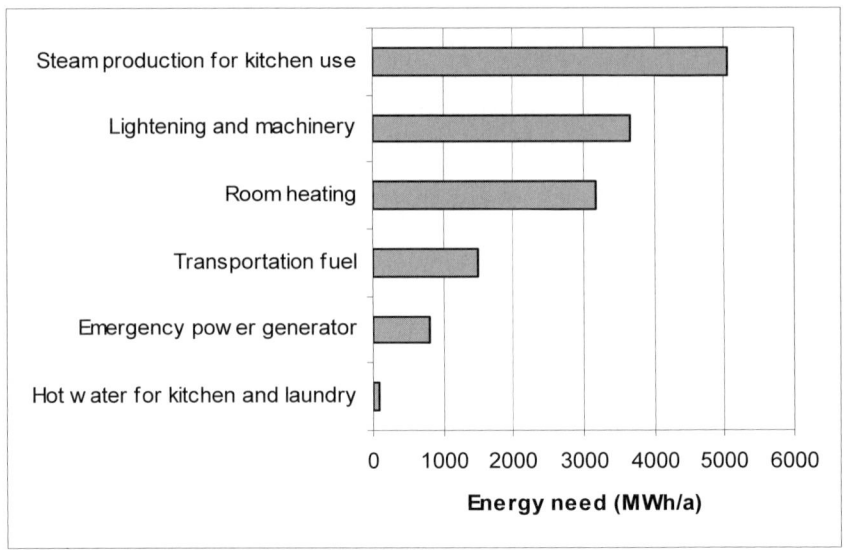

Figure 11.4 Profile of energy need in a recreation centre

The institution operates an animal farm and agricultural crop production and has a waste water treatment plant. All residues including manure, feed residues, sewage sludge, are at present mostly disposed on a landfill. They are very well suitable as biogas sources. The range of such substrates includes about 650 t/a cattle and pig manure, 6,000 t/a sewage sludge and kitchen leftover. Moreover 1,500 t and 370 t corn silage and pressed grain, respectively, not used for human nutrition, are available for biogas production. The GHG emission potential of both cattle and pig manure would be about 1,800 t CO_2-eq. annually if left untreated.

For the implementation of a biogas plant and the use of the resulting gas there are several options:

In a **basic scenario** all available substrates are used directly for the biogas production. Biogas then is utilized by combined heat and power (CHP) production. The scenario can, through variants of the economic conditions (such as cost reduction for construction, and usage of multiple generators) and addition of substrates such as extra crops or substrates from external animal farming, be changed. A cut down version of a biogas plant without block heater is taken into account as a **second scenario** with the thought of reducing total construction costs of the plant itself, since the main costs of building the plant are around the installation and maintenance of the block heater. In this case biogas would be an alternative to using coal for running the steam generator or produce warm water for room heating. This would avoid GHG emissions from coal burning and have a positive affect on environmental values regarding to smoke and ash fallout. Using biogas as a direct fuel for vehicles is a **third scenario** and is becoming an international trend. A precondition is that the gas must first be purified and compressed, for which a variety of equipment is available on the market and already in use. A supply into the natural gas net is also possible and is becoming economically reasonable.

The results of the calculations indicate that a biogas plant consisting of two reactors and a capacity of 2,000 m^3 is technically feasible under the given substrate supply potential. The electrical efficiency is 250 kW; also 750 kW heat is produced. An amount of 9,000 m^3 nitrogen rich residues can be irrigated as a fertilizer in agriculture.

Amongst the variants the use of biogas as a substitute for gasoline and the preparation of warm water for room heating are both economically reasonable at recent fuel costs. In the first case an amount of 1,400 l gasoline which is equivalent to 15,000 km driving distance is substituted. In the second case the total room heating expense can be covered. The GHG avoidance potential for coal and electricity substitution amount 1,000 and 3,000 t CO_2-eq./a, respectively, and would be 1,800 t more if the prevention of emission occurring from untreated manure would be taken into account.

11.4 Rice Cultivation

Rice is the most important cereal for the world's nutrition with an annual production capacity of about 620 Mio t (FAO, 2005). 95 percent is cultivated in China, India, and South East Asia. However the largest part is for consumption (subsistence farming) in the countries where it is produced. Only five percent of the world's rice production enter the world market. The most important exporters are Thailand, the USA, and some European countries, especially Italy, Spain, and France.

Rice cultivation heavily contributes to climate change by emissions of methane during the cultivation. Total methane emissions on a global scale are estimated 30 to 50 Mio t CH_4 annually which is equal to 750 to 1,250 Mio t CO_2-eq./a. In the U.S. agriculture rice cultivation emissions amount 400 Mio t CO_2-eq. or 1.3 percent of the country's total greenhouse gas emissions from agriculture.

Methane emissions are characteristic of the so-called wet system which applies to about 75 percent of the total rice production. Preconditions for the wet cultivation method are sufficient amounts of water as well as a warm and humid climate. Seedlings are positioned in flooded fields, for a first crop. In many countries after the first crop is harvested a second crop from the regrowth of the stubbles, so called ratoon, is possible. The specific methane emissions factors after field experiments in rice paddies are, averaged over a broad variety, between 210 and 780 kg methane per hectare and season for first and second crops, respectively.

The methane emission mechanism is as follows: In flooded systems organic material is decomposed by microbial activities in the soil and in the floodwater. After the oxygen is exhausted, anaerobic degradation results in the production of methane. The more organic matter is available to decompose the more methane is produced. However, the largest portion, about 60 to 90 percent, is oxidized by aerobic methanotrophic bacteria in the soil and thus is metabolized to CO_2 which is emitted to the atmosphere. Only are transported minor amounts of methane to the surface of the cultivation area and are emitted into the atmosphere: Some parts bubble through the water or are transported to the surface by diffusion. Some methane is transported through the rice plants and is emitted directly from the soil into the atmosphere. This path stands for about 90 percent of the methane which is transported to the surface. Another part is dissolved in the water and leaves the cultivation area by flushing when water is exchanged and is emitted afterwards.

The net emission of methane in the wet system depends on the rice water management applied which therefore implies the most effective mitigation potentials. Most is emitted from shallow systems. Deep water rice with a water depth of more than one meter is characterized by much lower emissions, due to the fact that the lower stems and roots of the plants are dead so that the movement of the methane through the plants is inhibited. If the field is drained periodically and the soil is dried sufficiently, methane is no longer produced and thus not emitted. Fertilizer practices also influence the methane production, especially the application of organic fertilizers which enhance the amount of decomposible matter. Some fertilizers such as nitrate and sulfate fertilizers inhibit the methane production.

Emission potentials are different between primary and secondary crops with secondary crop emissions nearly fourfold the value of primary.

As another established system the dry cultivation is operated without flooding the rice fields. No massive methane emissions occur. A shift to the dry system is therefore seen as a big chance to reduce the greenhouse effects of rice plantation. However this would need changes in technology and agricultural structures which only can be realized over a long term and is an expensive process. In some regions with high rainfall the dry system cannot be established at all. Changes in the traditional wet cultivation seem a more realistic mitigation approach in the shorter term.

11.5 Agricultural Soils

Agricultural soil contributes to the greenhouse effect, mostly due to nitrous gas (N_2O) emissions. A minor, but positive, effect is caused by methane consumption in soils.

N_2O is emitted from every type of soil as a result of naturally occurring microbial nitrification and denitrification. Both are parts of the natural nitrogen cycle. Nitrification is the aerobic microbial oxidation of ammonium (NH_4) which is a result of the decomposition of organic matter in soil, into nitrate (NO_3). Denitrification takes place under anaerobic conditions. Microorganisms then reduce nitrate to nitrogen gas (N_2). N_2O is an intermediate during both nitrification and denitrification, where the process of denitrification is well understood and is known to predominate.

The amount of N_2O produced during these processes depends on agricultural activities and the amount of nitrogen available in the soil. The predominant factor in enhancing the nitrogen level is the application of fertilizers, such as mineral nitrogen or organic fertilizer from livestock manure, compost, the working of plant residues into the soil, or the direct application of sewage sludge. Also nitrogen fixing crops, such as leguminoses, contribute to the N-level in soils. Indirect N_2O emissions from agriculture come from leaching and surface run-off from fertilized areas.

In U.S. agriculture N_2O from soil was 365 Mio t of CO_2-eq. per year in 2005 (EPA, 2007b). Its contribution to the overall agricultural greenhouse gas emissions is about 68 percent (without CO_2). This emission is about 75 percent of the nitrous gases in the national GHG balance of the United States, or 5 percent of the total national GHG emissions (in 2005; EPA, 2007b). In EU-15 direct soil emissions are 29 percent of the total N_2O emissions (EC, 2007). For German agriculture N_2O from soil management activities comes up with about 55 percent of total agricultural GHG emissions (without CO_2). A share of about 30 percent can be allocated to the use of fertilizers in the soil. 25 percent consist of indirect emissions resulting from leaching. 15 percent are due to spreading of farm manure or are from cultivated organic soils. The remaining third results from grazing, legumes, crop residues and from deposition of reactive N species which cause indirect GHG emissions (UBA, 2007).

A small positive effect of soil activities on climate results from consumption of atmospheric methane that is oxidised by methanotrophic bacteria which exist in soil. After European measurements CH_4 consumption is between 1.5 and 2.5 kg per hectare per year for farmland and grassland, respectively (UBA, 2007). In Germany soil is estimated a sink of 0.6 Mio t CO_2-eq., or about one percent of total agricultural GHG emissions (without CO_2).

12 Climate Effects of Waste Management

12.1 Background

Municipal solid waste is the end product of the life cycle of all solid material consumed. Its management is a big environmental challenge facing all countries since it influences material use and the depletion of natural resources, as well as the need for landfill space, and health problems. The direct impacts on climate are relatively small (less than 5 percent – see figure 9.1).

Greenhouse gas emissions occur at every step of the life cycle of material which is finally transformed into waste. This comprises the extraction and processing of the raw materials, the production of goods and services, the transportation of the raw materials and of the products to markets and to consumers, as well as the waste management after a product or a material becomes a waste. Waste management decisions can influence each of these steps. Strategies such as green design (EU, 2005) are important measures to reduce negative climate impact.

Different strategies were established, which may be, according to their potential to reduce climate impact, ranked as follows (see also figure 8.3):

1. Waste avoidance and source reduction
2. Reuse and recycling of waste, including composting
3. Waste pre-treatment before deposition, including waste stabilisation by biological methods and waste combustion
4. Ecologically sound disposal of residual waste in landfills.

All waste management activities provide opportunities for reducing GHG emissions. Waste avoidance and source reduction as well as recycling are often the most advantageous practices in waste management. Which degree of reduction is achieved depends on the individual circumstances, amongst which the composition of waste dominates. Moreover, the specific technology applied influences the calculation. Therefore only a material and energy specific comparison of all options exactly defines where the benefits are biggest. An example for such a decision making process using LCA is given in chapter 8.3.

In the following chapter, typical climate effects of some waste management activities are considered. They may help to decide what should be planned and realised in a concrete situation.

Some examples may give a first rough indication of the possible effects on various economic levels – see table 12.1. For more detailed calculations see chapter 12.7.

Table 12.1   Waste management action effects

Action level	Measure and climate effect
Company	Recycling of 50 t of paper and 4 t of aluminum per year instead of deposition reduces the GHG emissions by about 350 t CO_2-eq. (EPA, 2006).
	Re-use of carbon dioxide from composting of 1,000 t of green waste substitutes 150 t of technically produced CO_2 used as a fertilizer in a greenhouse (EPA, 2006).
	Reducing of plastics by 38 t and avoidance of 266 t metal saves 613 t CO_2-eq. in embedded energy (Reckitt, 2007).
	Replacing plastic blister packs with recycled and recyclable cardboard packaging saves 680 t plastic packing, equal to 2,430 t CO_2-eq. per year in embedded energy (Reckitt, 2007).
Community	Increase of the recycling rate from 30 to 40 percent at an average waste generation of one kg per person per day in a community of 30,000 and disposal at a landfill without a gas collection system results in a reduction of GHG emissions by 10,000 t CO_2-eq./a (EPA, 2006).
	In a town of 50,000 with a waste mass of 30,000 t per year the installation of a landfill gas recovery system reduces emissions by 22,000 t CO_2-eq./a (EPA, 2006).
	By recycling all of one family home's waste newsprint, cardboard, glass, metal, and organics, carbon dioxide emissions can be reduced by about 500 kg CO_2-eq. annually (EPA, 2006).
City (1 Mio)	Waste management in a mass burn combustor unit instead of deposition on a landfill without gas collection reduces GHG emissions by about 450,000 t CO_2-eq. (EPA, 2006).
Nation	On the national level in the USA an increase of the average recycling rate from 30 to 35 percent reduces GHG equivalent to 10 Mio t CO_2-eq./a (EPA, 2006).
	Current U.S. recycling efforts reduce greenhouse gas emissions by 49.9 Mio t CO_2-eq./a which is equivalent to the annual GHG emissions from 39.6 million passenger cars (UNFCCC, 2007).
	In Germany a shift from deposition of all MSW waste to combustion reduces total national GHG emissions by 0.4 percent (see also figure 8.4).

12.2 Source Reduction and Waste Recycling

12.2.1 Background and preconditions

Source reduction and waste recycling are two important options to improve waste management which is second in the range after waste avoidance.

In the case of "source reduction", less material is used to produce a product. This is achieved by practices like "green design" or "ecological design". Such activities are targets of the EU strategy for an improved resource management (EU, 2005). It is also possible to source reduce one type of material by a substitute which consists of another type of material with lower GHG emissions.

In the case of recycling the material is used in place of a virgin input in the manufacturing process, instead of being disposed and managed as waste. The material after its first use is recovered and prepared for a second use in the same field of application. Examples are the paper recycling or the use of retread tires. In a "closed loop" recycling the material is used to produce new material of the same kind, for example newspapers which are recycled into new newspapers. However, most of the material is recycled into a broader variety of manufactured products. This type of recycling is named as "open loop".

Benefits of recycling due to GHG emission reduction are calculated as the difference between GHG emissions when manufacturing from recycled or unhandled virgin material only.

New fields of application of waste components may also be opened for recycling if a suitable physical, chemical, or biological treatment of the original waste is applied. An example is the granulation of used tires or of plastic waste for a second use as a filling material in construction.

A third kind of recycling is oriented towards the processing of wastes into basic chemicals and their use in production processes. Examples are the gasification of plastics components for the production of methanol or the use of scrap metals from old cars in steel manufacture.

For basic information on these waste management activities and their influence on climate first a typical waste composition should be considered, which comprises the waste components most likely to have the greatest impact in GHGs. Such a list is given in table 12.2 in the case of the USA and comprises two-thirds of the waste which can be considered most important from the aspect of quantity generated, the potential contribution to methane production if deposited in a landfill as well as with respect to the difference of energy and material use for manufacturing from virgin or from recycled material.

Table 12.2   **Most decisive components of MSW (USA, 2004) (EPA, 2006)**

Material	Mass percent
Corrugated cardboard	13.0
Yard trimmings	12.0
Food discards	11.2
Newspapers	6.5
Glass	5.5
Dimensional lumber	3.4
Magazines/Third class mail	3.3
Office paper	3.2
High density polyethylene (HDPE)	1.6
Low density polyethylene (LDPE)	1.3
Steel cans	1.1
PET	0.8
Aluminum cans	0.7
Textbooks	0.5
Phonebooks	0.3

12.2.2 GHG effects of source reduction and recycling

By means of source reduction or recycling such GHG emissions are avoided which are caused by making the material and managing the post-consumer waste. Manufacturing from recycled material requires less energy, so that lower GHG emissions occur compared to manufacturing from virgin material. If waste cannot be avoided, source reduction is the most favourable GHG emissions avoidance method; for most materials it results in the lowest GHG values.

12.2.2.1 Source reduction effects

To estimate GHG effects of source reduction, the quality of the material to be source reduced must be known. Effects will be greatest if a product consists only of 100 percent virgin substances. However, in practice, a certain proportion of the material input is from recycled substances. Therefore, the effects will be smaller.

Source Reduction and Waste Recycling

The current mix of virgin and recycled inputs in the manufacture of selected poducts is given in figure 12.1.

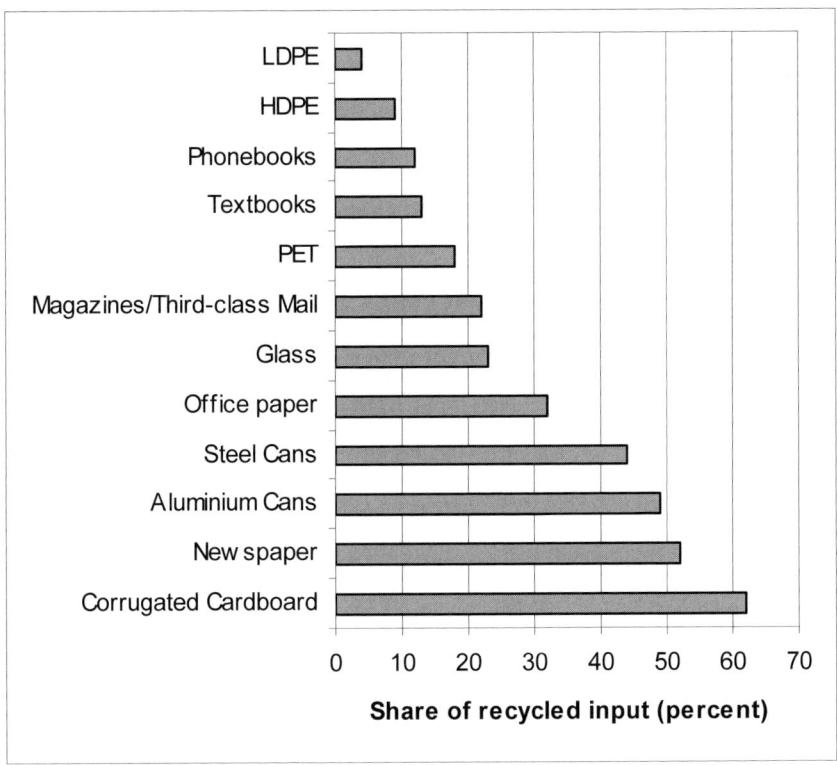

Figure 12.1 Current mix of recycled and virgin inputs of selected products (EPA, 2006c)

As figure 12.1 indicates, for corrugated cardboard, approximately two-thirds consist of recycled material. The portion of recycled paper in new paper products is in the wide range of between 10 and 50 percent. It strongly depends on the quality needs of the target product. Thus, the choice of the paper quality for a certain application also strongly influences GHG effects. In the case of aluminum or steel cans, nearly half of the raw material used is recycled. The portion of recycled plastics is relatively low.

Greenhouse gas effects by source reduction are given in figure 12.2 for selected materials for the two cases discussed: In "source reduction (virgin)" material was prepared only of virgin material. In "source reduction (mix)" the current mix of virgin and of recycled material is considered. As a third column the case of recycling instead of source reduction is presented – for details see later.

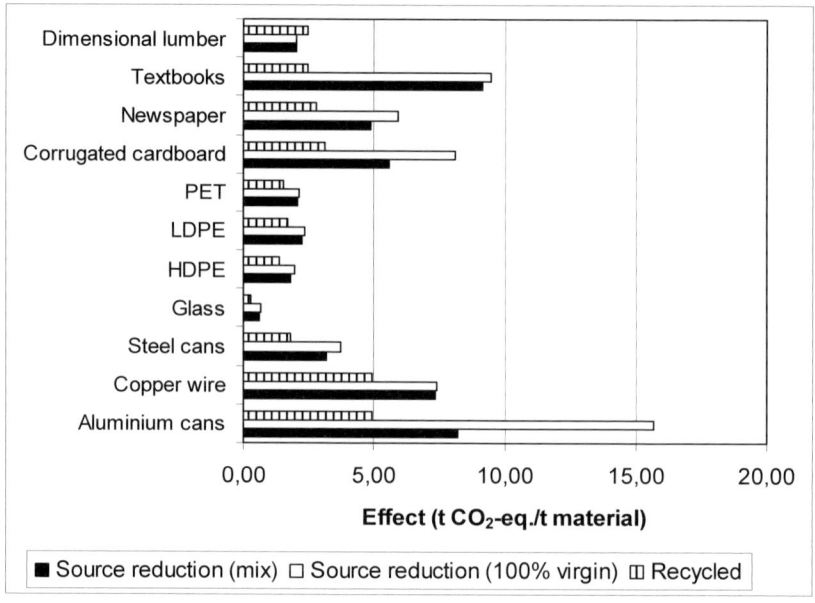

Figure 12.2 GHG effects of source reduction and recycling (EPA, 2006c)

Obviously, credits by source reduction using virgin material are always higher than for mixed inputs. The difference depends on production emissions. In the case of aluminum for which the emission reduction is highest amongst the materials displayed in figure 12.2 GHG effect of source reduction of 100 percent virgin material is about twice the value of mixed material due to high energy input into aluminum production (see also chapter 10.2.4.). In practical cases source reduction of aluminum will result in about 8 t CO_2-eq. per ton of material. Other important source reduction effects result with copper and specific paper grades. Source reduction activities therefore should first focus on these materials.

12.2.2.2 Waste material recycling effects

Material that is recycled after first use is then substituted for 100 percent virgin inputs in the production of new products. Emissions are lower in the case of using recycled inputs rather than virgin inputs, which results in credits.

For the calculation of the credits loss rates during the whole process of collection of waste material, its processing, and for remanufacturing have to be considered: 100 percent recycling is not possible. Less than one mass unit of new material is made from one mass unit of the recovered material. Table 12.3

Source Reduction and Waste Recycling

displays typical loss rates for recovered material. Depending on material data are based on closed- and open-loop recycling.

Table 12.3   Loss rates for recovered material (EPA, 2006)

Material	Recovered material retained in the recovery stage (percent)	Product made (t/t of recycled inputs)	Loss rate	
			t product per t recovered material	kg lost per ton of recovered
Steel cans	100	0.98	0.98	20
Aluminum cans	100	0.93	0.93	70
Corrugated cardboard	100	0.93	0.93	70
Newspaper	95	0.94	0.90	100
Glass	90	0.98	0.88	120
Dimensional lumber	88	0.91	0.80	200
Medium-density fibreboard	88	0.91	0.80	200
HDPE	90	0.86	0.78	220
LDPE	90	0.86	0.78	220
PET	90	0.86	0.78	220
Phonebooks	95	0.71	0.68	320
Magazines/Third class mail	95	0.71	0.67	330
Textbooks	95	0.69	0.66	340
Office paper	91	0.66	0.60	400

Calculations of climate effects of recycling based on these assumptions are shown in figure 12.2 in terms of the reduction of greenhouse gas emissions. The numbers characterize the improvement of emissions due to a waste generation reference point which is defined as the situation when the material has already undergone the acquisition of the raw material and the manufacturing phase. For more GHG reduction dates from selected material see table 12.14. Figure 12.2 indicates that for all the materials considered a reduction of greenhouse gas emissions would occur if a source reduction or a recycling takes place. Again, as was true in the case of source reduction, the greatest potential for emission reduction applies in the case of aluminum cans and several paper grades. Thus, if such measures are intended, they should start with these materials, depending on the concrete waste management situation.

Emission reductions caused by recycling activities are due to several factors, which contribute to total GHG reductions, namely the process energy, transportation energy as well as process emissions which are not energy related. Figure 12.3 represents recycled input credits for selected materials.

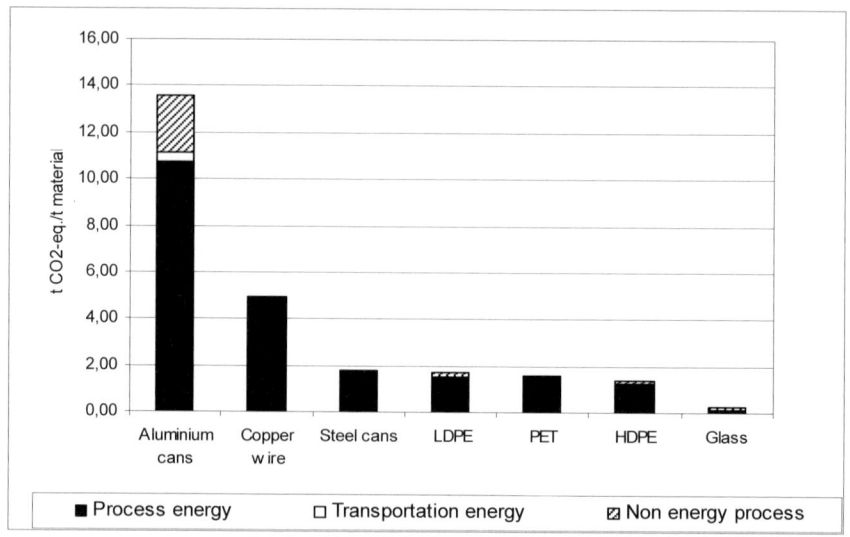

Figure 12.3 Effect of recycling process steps on emission reduction (EPA, 2006c)

In most cases the credits for the reduction of the process energy related emissions dominate. In the case of aluminum this amounts about 11 t CO_2-eq. per ton of recycled material used instead of virgin material. For aluminum also emissions from the process itself are a relevant factor. In the case of paper and products made from it positive recycling effects are also due to forest carbon sequestration which amounts up to 2 t CO_2-eq. per ton of wood (EPA, 2006).

12.2.3 Case study: GHG effects of the German packaging material recycling system DSD

As a result of the European Packaging Directive manufacturers and distributors of packaging are obliged to take back and reuse, recover or recycle the packaging they have put onto the market. But it is not necessary to do it themselves. In Germany they may be exempted from their obligations by participating in a system which collects sales packaging from consumers on a nationwide scale. Participation in such a system must be indicated by marking the packaging in question, and evidence of participation must be submitted to the competent authorities on demand (Hagengut, 2002).

In Germany a private organisation named DSD (Duales System Deutschland AG) was founded in 1990 by the business community for the retail trade, the consumer goods industry and the packaging industry. It operates as a non-profit organisation. A nationwide collection system for sales packaging was set up. Services are provided to almost 100 percent of households. The system is financed by the trade mark „Grüner Punkt (Green Dot)".

Under this system manufacturers apply for DSD and pay for corporation a fee to place their symbol of DSD, the Green Dot, on their packages. DSD collects and recycles the packages, instead of the producers of the packaging material which are responsible for it by law. The Green Dot shows the consumer that the package can be put into separate bins or sacks which exist in most households for collection purposes. In 2000, according to mass flow verification, nearly 5.67 Mio t of used sales packaging were collected. 96.5 percent were forwarded for recycling (Hagengut, 2002). In 2006 approximately 0.6 Mio t of plastic sales packaging were successfully recycled into regranulates, and thus made into new plastic products. The total GHG emission reduction was an estimated equivalent of 1.7 Mio t CO_2 (DSD, 2007).

The material collected is treated in sorting plants of which currently about 250 exist in Germany. The treatment involves a variety of different steps, such as dry mechanical pre-sorting, wet mechanical preparation, and plastic processing. What results is about 80 percent of secondary raw material and a residue consisting of wood, textiles, and stones. The composition and the material budget are given in table 12.4.

Table 12.4   Secondary material and residues after DSD (2002)

Component	Percentage
Tinplate	23.5
Beverage cartons	5.0
Paper fibres	8.0
Aluminum	4.0
Polyethylene granulate	13.0
Polystyrene granulate	3.5
PET	1.5
Poly-olefine agglomerates	23.0
Residues (wood, textiles, minerals)	18.5

Options other than the DSD Green Dot System are under discussion for the collection and treatment of light weight packaging amongst which only the Red Dot system involves the collection of large plastic packaging (see table 12.5).

Table 12.5 Differences in waste management options for packaging material (DSD, 2002b)

System	Collection	Recycling / Disposal
Green Dot	Collecting of lightweight packaging in a kerbside system within easy reach of households; residual waste in grey bins	Automatic sorting of lightweight packaging fraction by materials (so-called SORTEC technology) with subsequent re-processing of all materials; 100 percent high quality mechanical recycling; 100 percent combustion of residual waste
Red Dot	Reduced collection of lightweight packaging (only for large plastic packaging) via container bring system; residual waste and small plastic packaging in grey bins	Mechanical recycling of plastics; residual waste combusted
Combustion	Collecting of lightweight packaging together with residual waste via grey residual waste bin	Combustion with 50 percent energy use for electricity, steam, district heating; scrap reprocessing from slag

The total climate impacts of these systems may be calculated as the difference between

- expense in energy and material for the establishment of the system, the collection and the treatment of packaging material
- benefits from avoided process and energy needs for the material collected.

The budgeting of these items results according to table 12.6 (DSD, 2002b), which displays GHG effects and other eco-balancing characteristics.

Table 12.6 Results of environmental balancing for 2 Mio t lightweight packaging

Environmental Indicator	Unit	Green Dot	Red Dot	Waste combustion
Greenhouse effect	Mio t CO_2-eq.	-1.7	0.11	0.61
Acidification potential	10^3 t SO_2-eq.	-12	-2.1	-0.25

Environmental Indicator	Unit	Green Dot	Red Dot	Waste combustion
Nutrition potential	t PO$_3$-eq.	-1,400	-320	-92
Energy need	PJ	-49	-30	-25

The results indicate that collecting and treatment of packaging waste by the DSD system is not a burden on the environment at all. With respect to greenhouse effects the application of the Green Dot system reduces the CO_2 burdens by nearly 2 Mio t CO_2-eq. compared to a system without collection of packaging material. Also with other environmental indicators, such as acidification, nutrition, and energy, positive environmental effects may be achieved.

The conclusion is that the more packaging waste is collected the better. Careful waste separation at home means a tangible contribution from each private individual to resource economy and to climate protection.

12.3 Composting

12.3.1 Composting process characters

Composting is a technology for the treatment of organic residues using aerobic bioprocesses. Organic material, which consists of sugar, starch, cellulose, hemi-cellulose, and a lignin like fraction, is fully or partly decomposed by different kinds of micro-organisms which act in a complicated metabolic pathway. The result of the composting process is compost. It mainly consists of those organic waste components which are not or only partly used by the microbial metabolism, as well as of components which are formed in the longer term during the so-called maturation processes.

The compost is used as fertilizer in agriculture. Benefits arise from the nutrient content of the compost, like salts of potassium, phosphorus, and nitrate. But it is even more important that the organic matter in the compost, such as humus like substances, improve the concentration of organic matter in the soil and its structure, and preserve soil fertility over a long period.

Sources of compost are wastes from agriculture such as crop residues, wastes from gardening, yard trimmings, as well as source separated kitchen waste. In Germany a capacity of about 4 Mio t of separately collected biowaste is treated and processed into compost every year.

The technology of composting comprises different methods such as open windrow systems as the simplest, and closed reactors as the most sophisticated technology. Composting reactors are characterized with high process intensity due to a good aeration capacity, by which anaerobic processes are largely prevented. This is also a pre-condition for good compost quality.

Besides industrial composting in centralised facilities of capacities up to 100,000 t a year, home composting of garden residues takes place, mostly in open heaps (see figure 12.4).

12.3.2 GHG sources in composting

Composting may result in emissions from various sources, such as

- biogenic processes during composting,
- process gas cleaning and process control,
- collection and transportation of the raw material and the compost,
- the application of compost in agriculture.

Main gas components to be considered are CO_2, CH_4, N_2O, and NH_3. A qualitative review of the emissions includes the following emission types:

- Emissions from the process itself mainly consist of carbon dioxide which is the result of the aerobic decomposition. Depending on the type of raw material, the duration of the composting process, as well as other bioprocess characteristics, different amounts of CO_2 are emitted per ton of composted raw material. Because CO_2 in this case is biogenic in origin, this emission is not counted in greenhouse gas inventories. Nevertheless capturing of emitted CO_2 and its use instead of carbon dioxide from fossil sources will improve the anthropogenic greenhouse gas balance (see chapter 12.3.4).

- In a well-managed composting process, CO_2 is the only process gas. If aeration in the compost heap is poor, or the material is too wet, an anaerobic situation may occur, which is accompanied by methane development and the liberation of emissions of odor. Emission factors of methane are different for biowastes from households and from green wastes. The values are 2.5 and 3.36 kg methane per ton of biowaste treated, respectively (UBA, 2007).

- Nitrous oxide (N_2O) has to be taken into account. It results from the oxidation of ammonia which is another by-product of the composting process. Emission factors are different for biowaste from households and green wastes. The values are 83 and 60.3 g N_2O per ton of biowaste after experimental results in Germany. The total N_2O emissions from composting in Germany are about 0.25 Mio t CO_2-eq. or 0.02 percent of total GHG emissions (in 2004; UBA, 2007).

- Another source of N_2O are biofilters which are a component of composting facilities and aim to reduce or eliminate odors. They are applied also in other processes where organic emissions occur. N_2O in this case is the result of microbial conversion of ammonia. Therefore biofilters may act as climate gas sources if ammonia is not eliminated from the waste gas stream before entering the filter. As an example data of an experimental biofilter (in the case of MBP technology – see chapter 12.6.2) are displayed in table 12.7: N_2O concentration is raised from 19 to 130 g, measured as a specific amount per ton of waste. Such effects can be avoided if ammonia is eliminated from the waste gas stream by use of an acid washer and scrubber.

Table 12.7   **Nitrogen balance in a biofilter (g/t biowaste) (Soyez, 2001)**

Substance	Raw gas input	Clean gas after biofilter
N_2O	19	130
NO	1	190
NH_3	500	200
N_{org}	100	100
Percolation	0	<1-10

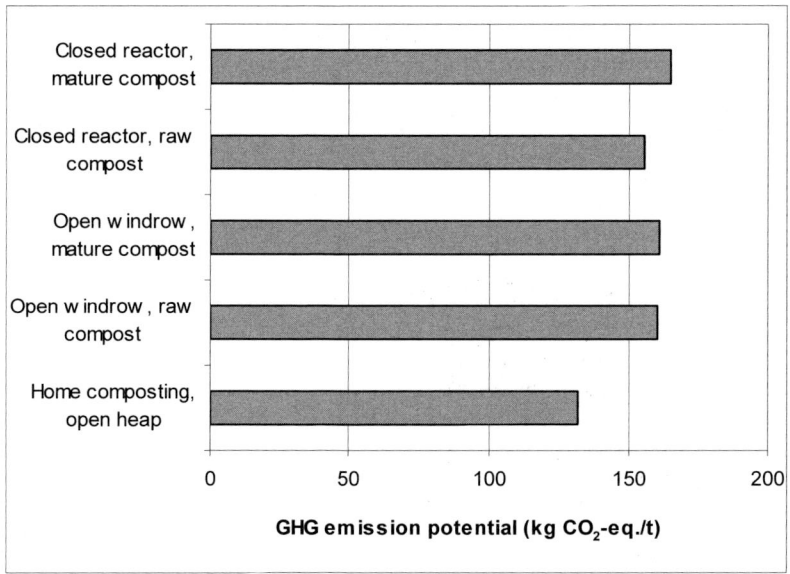

Figure 12.4   **GHG emissions by composting technologies (after Knappe, 2004)**

Besides direct process related emissions other technological steps contribute to GHG emissions. During collection of biowaste and its transportation to the composting facility, as well as during turning of compost and aeration, CO_2 and methane emissions take place.

The overall greenhouse gas emissions, both antropogen and natural, amount up to about 150 kg CO_2-eq. per ton of waste treated (see figure 12.4), depending on the technology and the type of the compost produced.

Obviously the emission values are quite similar for the technologies compared, with highest values for mature compost the production of which normally comprises an extra maturation step. Home composting represents the lowest value, since practically no energy consumption in transportation and handling is necessary. Thus home composting from climate perspective would be a favourable composting option if processed properly.

Additionally a third factor which is the application of compost as a soil fertilizer has to be taken into account. However though it results in GHG emission, it isn't counted in GHG balances due to its biological origin.

A specification of the contribution of the steps of processing and application is given in figure 12.5.

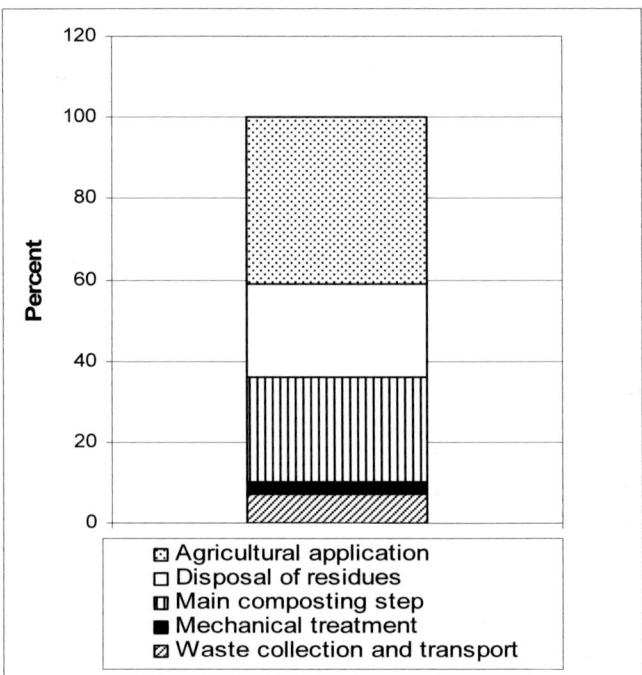

Figure 12.5 **GHG emissions of composting steps (after Knappe, 2004)**

Waste collection and mechanical treatment as a step of the composting process contribute by only 10 percent, the main process phase by not more than 25 percent. Largest effect is by agricultural application. If matured compost is produced this value is cut in half. However, in this case, the production effort is higher, so that the benefits are equalized and the total GHG effect is nearly unchanged. More than 40 percent is not counted in the GHG balance which is due to its biological origin.

12.3.3 Carbon sequestration by compost application

Compost is applied in agriculture to improve soil fertility by means of the supply of mineral fertilizers, such as potassium, phosphorus, and nitrogen. Moreover, the input of compost strongly influences the soil carbon storage which is also an important factor of soil fertility. This is due to the fact that composting partly results in the increased formation of stable carbon compounds, i.e. humus-like substances and aggregates. These are made of complex compounds that render them resistant to microbial attack.

The input of organic matter is especially important in such a case where an intensive cultivation of soil results in its degradation, since decomposition rates and removal of carbon by the crops are not well balanced by inputs. By adding compost an input of new organic matter takes place, so that the soil carbon level is restored. In this case compost nitrogen stimulates soil productivity which results in the higher volume of crop residues. Other compost components may have a multiplier effect, by which carbon mass accumulation is even higher than the direct carbon input by the organic compost mass.

The type of organic matter which is produced by composting can be stored in soil over many decades. Its decomposition rate has been estimated to be around 30 to 40 percent during the first year, and a decreasing rate later on. Field application of compost therefore is a temporary sink in carbon dioxide and results in a real net improvement of the overall greenhouse gas balance. The storage effect of soil carbon sequestration is in the range of about 0.054 (Smith, 2001) to 0.24 t CO_2-eq. per ton of compost applied (EPA, 2006). Taking into account a total of CO_2 emissions of about 0.150 t/t (see figure 12.4), in the case of a sequestration of 0.24 t/t there is a benefit for the total GHG balance of about 0.090 t CO_2-eq. per ton of compost applied.

On an EU level the use of compost from the biodegradable fraction of municipal waste is estimated to have a storing potential of 1.4 Mio t CO_2-eq. per year if the whole putriscible fraction of MSW is composted in all Member States (EC, 2004).

By application of compost, there are also other climate related benefits under discussion even if they are not easily measurable, such as i) improved workability of soils, which reduces the energy needs for machinery, ii) reduced erosion, which keeps more organic matter in the surface layer followed by higher crops, iii) improved water retention, which means less energy for irrigation, and iv) suppressive power against pests followed by reduction of pesticide related climate impacts (EUNOMIA, 2005).

12.3.4 Use of composting CO_2 as greenhouse fertilizer

If compost born CO_2 could be applied in production processes instead of fossil derived carbon dioxide, a net reduction of the GHG balance would be possible. As was mentioned a total of about 150 kg CO_2 is emitted per ton of compost raw material. Thus in a facility with a capacity of 100,000 t annually, about 15,000 t of carbon dioxide are produced. In Germany CO_2 from composting totals about one Mio t of carbon dioxide, which could be used instead of fossil derived CO_2 in industrial or related processes.

A sensible use of compost born carbon dioxide is its application in greenhouses where crops are fertilized by CO_2 which improves the yields by about 30-40 percent through a CO_2 input of 100 t per hectare annually. Conventionally CO_2 is from gas burners or is industrially produced. If compost CO_2 was used by a medium sized composting facility, an area of about 150 hectares could be fully supplied. As another advantage the residues from the greenhouse crop can be applied as raw material in the composting process. Moreover, renewable heat energy, produced by the composting process supports climatization of the greenhouse, hence avoiding climate gas emissions from fossil fuels (Soyez, 1990b). It is another advantage of such a combination that excess heat from the greenhouse could be used to support the composting process start, hence reducing energy needs.

12.4 Climate Effects of Waste Deposition in Landfills

Waste deposition in landfills is the final step in the waste management hierarchy.

The climatic effects from landfills mainly result from

- landfill gas emissions, especially methane, as a result of microbial decaying of putriscible matter in the waste deposited in the landfill,
- waste transportation and processing on the landfill site,
- carbon sequestration in the landfill body by forming stable carbon structures.

12.4.1 Climate Effects by landfill gas emissions

Landfill gas (LFG) in the year 2000 contributed to global methane emissions of about 13 percent with a total amount of 818 Mio t of CO_2-eq. per year (EPA, 2007g). It is third in the range of human induced methane sources. Nearly half of the total emissions stem from four countries: the USA, China, Russia, and Mexico (see table 12.8). More than a quarter of the world's total (26 percent) LFG emissions are emitted in the USA.

Table 12.8 Landfill methane emissions by the 10 most relevant countries in 2000 and estimations for 2020, in Mio t CO_2-eq./a (EPA, 2005)

Country	Reference year	
	2000	2020
USA	199.3	145.2
Russia	51.1	45.3
China	44.6	49.7
Mexico	31.0	39.2
Ukraine	25.2	37.4
Canada	22.8	33.5
Poland	17.0	17.0
Brazil	15.6	19.0
Australia	14.8	22.0
Germany	14.4	4.4

The global growth of landfill gas emissions is estimated to nearly 20 percent between 2005 and 2020, mostly due to the raising amounts of waste to be deposited under poor landfill management conditions in emerging countries and China, whereas in industrialized countries LFG emissions will decline due to strict regulations to reduce methane emissions. Examples of such regulations are the *U.S. Landfill Rule* (by 1999), or the *EU Landfill Directive* (by 2002). They principally include waste management improvements, such as

- waste reduction, re-use and recycling, e.g. the reduction of the organic content of waste by source separation of organics (see chapter 12.1),
- pre-treatment of the waste to reduce the organic content prior to deposition, e.g. by waste combustion or mechanical-biological pre-treatment (see chapter 12.6),

- the collection of landfill gas and use of its energy content in energy recovery systems for electricity production and in district heating.

The following examples illustrate the situation in Germany where requirements for landfills were imposed in 2001 by the *Ordinance on Waste Storage and on Landfills* (BMU, 2001). No further deposition will be permitted if waste has an organic content (measured as loss by incineration) of more than 5 percent. Hence, no waste with significant potential for methane formation will be deposited. For conformity with pertinent requirements MSW has to be pre-treated via thermal (combustion) or mechanical-biological processes. By this pre-treatment a reduction of the waste mass to be deposited is envisaged at 60 to 70 percent. Landfill gas in future will mostly originate from older landfills. As this tapers off, landfill methane emissions will decrease extensively and will reach less than 10 percent of the value of 1990 in year 2012 (UBA, 2007) or 10 Mio t less in 2020 compared to 2000 (see table 12.8).

12.4.1.1 Overview on landfill gas generation

The processes of landfill gas generation and emission are as follows: Organic compounds of waste such as paper, food discharges, or yard trimmings are decomposed just after being deposited in the landfill body. Initially they are metabolized by aerobic micro-organisms. This process lasts as long as oxygen is available in the waste mass, normally some months. After its depletion, anaerobic processes start which like the biogas process result in the landfill gas.

The typical dry composition of the low-energy content gas is 57 percent methane, 42 percent carbon dioxide, 0.5 percent nitrogen, 0.2 percent hydrogen, and 0.2 percent oxygen. In addition, a significant number of other compounds are found in trace quantities, especially NMVOCs by one percent at maximum. These include alkanes, aromatics, chlorocarbons, oxygenated compounds, other hydrocarbons, and sulfur dioxide. Some of these substances are climate relevant with a very high GHG potential (see table 7.1).

The gas generation starts about one to two years after waste disposal in the landfill and continues in significant amounts for some decades, depending on the composition of the waste. The magnitude of methane generation depends on the quantity, the type and the moisture content of the waste and the design and management practices on the landfill site.

The more organics which are contained in the waste the higher is the production of methane and thus the contribution of the waste to the greenhouse gas effect. Plastics, though organic in origin, are mostly not degraded by bioprocesses, with the exception of bioplastics. Metals do not directly contribute to the GHG emissions.

Some figures of landfill gas production by solid waste components which contain organics are given in table 12.9.

Table 12.9 Methane yield for solid waste components (EPA, 2006)

Waste component	Average measured methane yield (l/kg dry mass)	GHG effect (t CO_2-eq./wet ton)
Corrugated cardboard	152.3	1.969
Magazines/Third class mail	84.4	1.978
Newspaper	74.2	0.950
Office paper	217.3	4.426
Food discards	300.7	1.228
Grass	144.3	0.785
Leaves	30.5	0.609
Branches	62.6	0.623
Mixed municipal solid waste	92.0	1.049

Also other important factors with relevance to climate are influenced by the organic content, such as the production of leachates, and landfill settlings, which may cause instabilities of the landfill body.

12.4.1.2 Landfill gas recovery

A main factor in reducing methane emissions from the landfill body is to collect the landfill gas before it is released into the atmosphere. For this reason landfill gas recovery systems are applied: LFG is extracted from the landfill body using a series of wells. A vacuum system directs the collected gas to a point where it is processed.

Several types of processing are possible. In the case of a flare only the gas is burned so that its energy content is lost. Alternatively the gas can be used beneficially; this includes the use of the gas as fuel in energy recovery facilities, such as internal combustion engines, gas turbines, micro-turbines, steam boilers, or other facilities that use gas for power production. By this means up to an average of 70 percent of the gas generated can be captured and transformed into electricity and heat.

Besides direct avoidance of greenhouse gas emissions, LFG recovery prevents greenhouse gas emissions caused by the fossil fuel which would be needed for production of the equivalent of electricity and heat.

Some examples may illustrate the application of landfill gas recovery systems and its effects in the case of the USA, and of Mexico as a developing country:

- In the USA the national landfill gas budget accounts for 24 percent of all anthropogenic methane emissions; annual total was 126 Mio t CO_2-eq. by 2006 (EPA, 2008a). The emissions stem from 1,800 operational landfills. From 1990 to 2005 landfill gas emissions decreased by 18 percent. This downward trend is the result of increased gas collection and combustion intensity. In 2007, 435 landfill gas projects were active which had a capacity of 1,325 MW. These projects provided over 10.5 billion kWh of electricity, and delivered to corporate and government users landfill gas with an energy equivalent to powering roughly 800,000 homes and heating more than half a million homes each year (EPA, 2008). About 1,530 landfills with a capacity of 1,290 MW are candidates for landfill gas programmes coming into power in the next few years (EPA, 2007d). A recent example supported by EPA was a set of 22 landfill gas projects operated by a landfill company in New England. These projects will generate in excess of 110 MW of renewable energy (or equivalent). They remove from the atmosphere over 0.6 Mio t CO_2-eq. each year which is equivalent to planting in excess of 1.1 million acres of trees and offset the need for almost 21,000 rail cars of coal (EPA, 2008).

- In developing countries national as well as international programmes support the establishment of landfill gas recovery systems to reduce methane emission and use the energy for power production or household lightening. The following example is from Mexico, where landfills contribute to 10 percent of the total human-influenced greenhouse gas emissions. It demonstrates the possible effects (Simeprodeso, 2007): Beginning in 2001 in the city of Monterrey with nearly 4 million inhabitants, where over 4,500 t of municipal solid waste are disposed per day in the *Simeprodeso* landfill, methane from the landfill was harnessed for energy recovery while reducing methane emissions. A joint venture between government and business interests, which in part was funded by an US$ 5 Mio grant from the Global Environmental Facility, launched a project for transforming LFG into electricity. The energy is fed into the local net to help drive the public transit system by day, and light city streets by night, and to provide power for over 15,000 homes. It is planned to enhance the capacity of the existing power station, so that 80 percent of the municipal government's electricity needs will be met. Moreover, as the *Simeprodeso* landfill continues to expand, LFG generation is estimated to increase to fuel a 25 MW facility for completion by 2016.

12.4.1.3 Effects of landfill management

One-quarter of methane generated in the landfill cannot be captured by gas recovery systems due to low concentration of methane and resulting poor economy or is diffusely emitted during or after finishing the gas collection. A portion thereof can be oxidized in a landfill surface layer with a methane oxidizing substrate, such as compost or residues from mechanical-biological waste treatment. By this means the residual methane emissions can be reduced by about 60 percent in the case of the deposition of MSW.

In the case of deposition of waste pre-treated by the MBP technology (see chapter 12.6) emission characteristic is different: Emission potential is 10 to 45 Nm^3/ton of waste dry matter which equals about one-tenth of untreated waste. Methane production already starts after one month rather than one to two years in the case of untreated waste, and long lasting emissions at high concentration do not take place. Methane oxidation potential of a specially constructed landfill cover exhibits the landfill gas emission into the atmosphere. Hence practically no emissions will occur from the landfill after deposition of MBP pre-treated waste.

12.4.2 Carbon storage by solid waste deposition

Organic waste components, such as yard trimmings, food discards, and paper are not completely decomposed by the microbial processes, especially by anaerobic bacteria. Thus residues of organics which under normal natural conditions would be decomposed over time in the photosynthetic cycle remain in the landfill. Therefore the amount of carbon stored in the landfill body is considered to be an anthropogenic sink of carbon which reduces the burden of CO_2 emissions.

The amount of plastics which remain in the landfill also means storage of carbon. However, it is not counted as a sink, since it is of fossil origin.

Table 12.9 indicates the carbon storage efficiency of some selected organic residues in landfills.

Table 12.9 Carbon storage potential of waste components

Waste component	Amount of carbon stored (t CO_2-eq./wet ton)
Corrugated cardboard	0.81
Magazines/Third class mail	1.06

Waste component	Amount of carbon stored (t CO_2-eq./wet ton)
Newspaper	1.32
Office paper	0.15
Food discards	0.07
Grass	0.44
Leaves	1.43
Branches	0.77
Mixed MSW	0.37

However it is to be mentioned that an even better climate effect would be possible if the material was recycled instead of its deposition in a landfill (see figure 12.2).

12.5 Climate Effects of Waste Combustion

12.5.1 Technological background – Waste to Energy (WtE)

Combustion of waste is a traditional method of the treatment of waste prior to deposition. It results in a reduction of waste to be deposited by about 70 to 80 percent. Only the residues of the process, such as ash and slag, in an amount of about 20 to 30 percent, would have to be deposited if not used as construction material under certain environmental constraints. At present another target of waste combustion comes into the focus of technology development and economy: Instead of waste minimisation the energy recovery approach dominates whereby the energy content of the waste is used for the production of power and heat. Such combustion facilities are defined as waste-to-energy (WtE) plants.

The amount of waste incinerated recently was about 130 Mio t per year which is equivalent to an energy amount of more than 1 EJ assuming the energy content of 9 GJ/t waste. Since the global waste amount is estimated to contain 8 EJ (in 2002) and 13 EJ in 2030, a big potential for energy generation by combustion is envisaged (IPCC, 2007c).

WtE facilities comprise mass burners, modular plants, and refuse derived fuel (RDF) incinerators:

- Mass burners generate electricity and/or steam by combustion of mixed municipal solid waste which has an average heating value of 4 to 6 GJ per ton. The capacity of a single facility is in the range of several hundreds of kilotons per year. In Germany about 60 mass burners with a total capacity of 15 Mio t of municipal solid waste exist. In the USA about 70 facilities process about 34 Mio t annually.

- Modular WtE plants are similar to mass burners, using municipal solid waste (MSW), but are lower in capacity. They normally are prefabricated units and established on site, so that they are more flexible in application.

- RDF facilities are specially tailored to the treatment of a fuel which is derived from MSW by special processing. This includes separation steps to remove material with low or no heating value, as well as waste components containing fast decaying organics which potentially produce odors. One process option is MBP (see chapter 12.6). The resulting fraction is of high calorific value, typically in the range of 15 GJ/t. It is more uniform than MSW and can be stored up to one year.

12.5.2 Climate effects by MSW combustion

Combustion is a chemical process where components are oxidized by deliberation of reaction energy in the form of heat. In the case of waste combustion, oxygen normally is supplied by air which in the process is transferred into a waste gas. It contains the combustion products such as CO_2, N_2O and other oxidation products, as well as non combusted gaseous residues from the original waste. The typical amount of waste gas produced is about 5,700 Nm^3 per ton of waste (BREF, 2006).

Climate effects by waste combustion result from both

- direct emissions of CO_2, N_2O, pollutants, and indirect emissions
- avoiding greenhouse gases by energy and material recovery

Net greenhouse gas emissions result from the addition of these effects.

12.5.2.1 CO_2, N_2O and pollutant emissions

During combustion nearly all carbon substances of the waste are transferred into CO_2. The specific climate effect of CO_2 from combustion is about 0.7 t CO_2-eq./t MSW. CO_2 from the biogenic carbon in the waste, which is about 60 percent (see table 10.7), is not counted as a greenhouse gas. Only N_2O contributes to about two percent. In the USA where about 34 Mio t of MSW were combusted in 2005, a total of 21.3 Mio t CO_2-eq. was emitted from which 20.9 and 0.4 were from CO_2 and N_2O, respectively.

Besides CO_2 a broad variety of pollutants are emitted. Some of them are climate relevant, others are indirect climate gases. A selection of such substances is given in table 12.10.

Table 12.10   Emissions caused by waste combustion

Pollutant	Emission load (mg/m³ waste gas)
Anorganic substances	
SO_2	6.72
NO_x	111
CO	25.4
NH_3	5.4
Organic substances	
Dichlormethane	0.014
Hexachlorbenzene	0.000010
Polychlorated biphenyls	0.000015
Trichlormethane	0.000900

Other indirect emissions occur from supporting processes such as fuel supply in the case of poor heating value of the waste combusted, and natural gas in an amount of 11 m³ per ton of waste for waste gas treatment, as well as for transport operations in the facility.

12.5.2.2   Beneficial climate effects by recovery

Recovery activities refer to electricity and steam production as well as material recovery.

Electricity production is a result of the combustion in facilities with energy recovery such as WtE and RDF facilities. The benefits on climate of this process result from the fact that CO_2 emissions which would otherwise be provided by an electricity utility power plant which burns fossil fuels are avoided.

To which extent climate impacts can be reduced depends on the GHG impacts of the energy substituted. A more generalised GHG avoidance number will result if the average mix of fuels in a regional or a national economy is considered, which value depends on the proportion of power sources (see 13.1.1). In the case of Germany or the USA it amounts to 0.0549 and 0.0794 t CO_2-eq. per GJ, respectively. In this case for every ton of waste which is combusted for power production instead of the mix given GHG impacts of 0.8 to 1.2 t CO_2-eq. are avoided.

Avoidance of GHG emissions at material production also takes place, since most MSW combustion facilities recover ferrous and non-ferrous metals and glass. Benefits of recovery of recyclable material can be estimated after the concrete amounts separated and the emission factors given in chapter 12.2. They are similar to effects by MBP technology which is discussed in more detail in chapter 12.6.

12.6 Climate Effects of Mechanical-biological Waste Pre-treatment

12.6.1 Technological background

Mechanical-biological waste pre-treatment (MBP) technology is a very recent option in waste management. The first facilities started operation in the mid-1990s. MBP comprises the processing or conversion of waste from human settlements, and waste that can be managed like waste from human settlements, with biodegradable organic components, via a combination of mechanical and other physical processes (e.g. cutting or crushing, sorting) with biological processes ("rotting" or "decomposition", fermentation), on which

- biologically stabilized waste is produced for deposition or prior to thermal treatment,
- thermally valuable components or substitute fuels (refuse derived fuels, RDF) are obtained for recovery, or
- biogas is generated for energy recovery (BMU, 2001).

The technology is now being widely implemented in Europe, especially in Germany and Austria. In Germany, in 2006 about 3.8 Mio t of residual waste were operated in 46 MBP facilities. The total treatment in Europe was about 13 Mio t. In developing and emerging countries, such as China, Vietnam, and Brazil, MBP projects were built to improve the waste management situation (Nelles, 2007).

The typical process is characterized by mechanical separation steps to produce i) the RDF fraction which typically amounts to about 40 percent of the original waste. ii) 35 percent by mass of a fine fraction is produced which is rich in organics, but has a low heating value. After separation, this fraction is biologically treated to reduce the organic content to a low value so that MBP treated waste can be deposited at low risk of emissions (Soyez, 2001). The residual organics must not exceed a certain level for the respiration activity ($AT_4 < 5$ mg/g) or alternatively gas formation in a laboratory test (GB_{21} <20 l/kg). The upper thermal value of the waste deposited must be lower than 6,000 kJ per kg. Thus an energy rich fraction is separated.

A typical process scheme is given in figure 12.6.

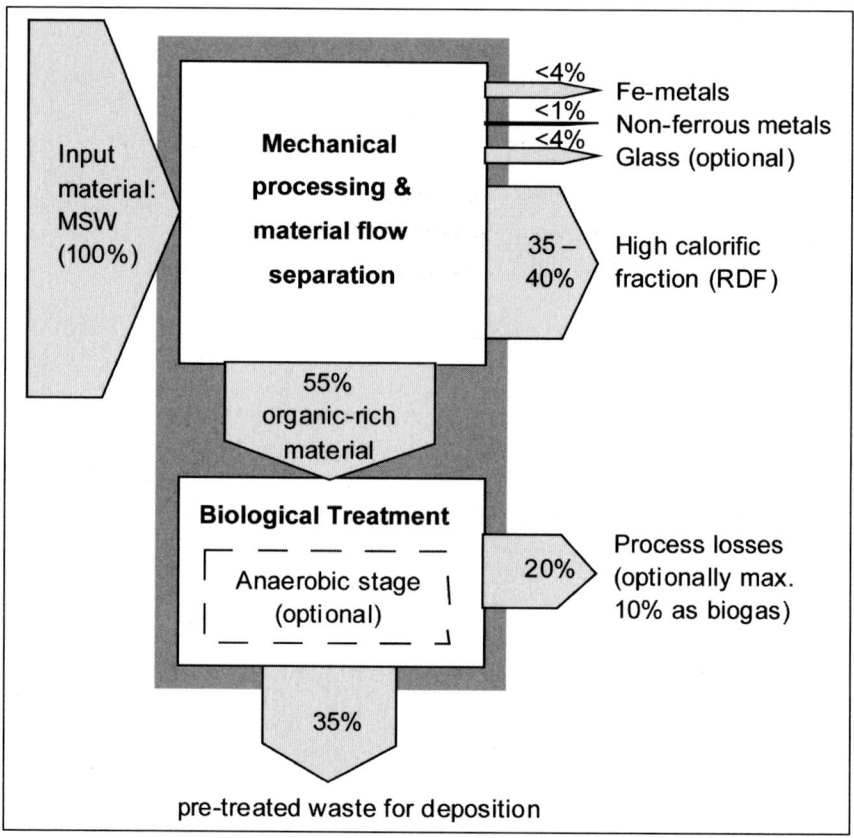

Figure 12.6   **Process schema of RDF production by MBP (Soyez, 2001)**

Two types of processes for the biological treatment or combinations thereof can be applied according to the MBP technology: aerobic and anaerobic processes. In the case of aerobic processes the organic residues are oxidized. This is similar to composting. However the product is not "compost" because its quality is poor due to heavy metals and toxic substances. The energy content of the waste in this case is lost. In case the biological treatment is executed by an anaerobic process, methane will be produced and transformed into electricity to cover the needs of the process itself or for energy marketing. Waste heat can be used for district heating or for technical processes such as drying.

As another important aspect, in MBP facilities a material recovery takes place. Especially ferrous and non-ferrous metals are recovered by about four and one percent, respectively. Optionally a recovery of glass is possible and is applied depending on the market needs. The recovery of these materials avoids climate impacts in the production processes and is thus considered as a climate benefit of waste pre-treatment.

There are MBP variants established which only produce RDF and recyclable materials such as metals and glass, and no waste for deposition. The plant type may be directly combined with a combustion unit to produce energy, or it produces an RDF fraction for marketing in the energy sector as well as for cement or methanol production. A MBP facility situated in Dresden (Germany) is shown in figure Figure 12.7.

Figure 12.7 **Total view of an industrial scale MBP facility in Germany**

The output flow of such a process is characterized by about two-thirds of the fractions for industrial re-use; only 4 percent are deposited, relevant to 100 percent input, not including process losses (see table 12.11).

Table 12.11   Output flow composition of MBP technology (Soyez, 2002)

Output fraction	Percent of total input
Fraction for industrial re-use	
RDF (calorific value 15-18 MJ/kg)	53
Ferrous metals	4
Non-ferrous metals	1
Batteries	0.05
White glass	5
Brown glass	0.5
Green glass	0.5
Minerals	4
Others	
Fine grain and dust (to be deposited)	4

12.6.2   Climate effects by MBP technology

Climate effects caused by MBP technology can be addressed to several sources – see table 12.12.

Table 12.12   Climate impacts by MBP technology

Technological step	Climate impact source	Assessment of effects
MBP processing	Energy need of processes such as ventilation and transportation	Marginal, but negative
	Recovery of metals and glass	High CO_2 avoidance potential
	Emissions of process gases	Emission data: Allocation values by law (BMU, 2001): N_2O <100 g/t, TOC <55 g/t, Typical values (Doedens, 2007): N_2O = 6 g/t, TOC = 34 g/t

Climate Effects of Mechanical-biological Waste Pre-treatment   165

Technological step	Climate impact source	Assessment of effects
	Biogas production in an anaerobic process technology	100 m³/t waste
Deposition of pre-treated waste	Emissions from landfill	Negative due to GHG emissions, but improved compared to deposition of waste without treatment
Use of RDF fraction	Emissions by combustion in a WtE-facility	Positive mostly due to native organic biomass content, but also due to substitution of fossil fuel for energy production

12.6.2.1 Climate effects of material recovery

The ecological assessment for the most climate relevant LCA impact categories (see also chapter 8) of material recovery of iron, non-ferrous metals, and plastics, as well as of the RDF use in cement kilns is given in figure 12.8 as specific contribution to the national (German) GHG inventory. The calculation is based on the following typical range of parameters:

- The amount of metals in the residual waste varies between 2.5 and 10 mass percent. The share of ferrous metals is 75 percent or higher. After crushing and homogenisation about 95 percent of the ferrous metals are recovered by magnetic separators.

- The share of non-ferrous metals is 0.6 to 2.6 mass percent. It contains 30 to 50 percent of aluminum. The efficiency of separation is as high as 98 percent in case of an eddy current separator.

- In the case of anaerobic processing of organic residues in the MBP plant a gas yield of 25 to 125 m³/t MBP-Input is assumed. The gas produced has a methane content of 60 percent.

- The percentage of plastics within the residual waste is between 7 and 15 mass percent. Plastic consists of polyethylene only, of which 80 percent is recovered, and 70 percent of the recovered plastic is recyclable.

- The high calorific fraction has a heating value of 15,000 kJ per kg. It consists of plastics, paper, and packaging material. It is used as a secondary fuel in a cement kiln.

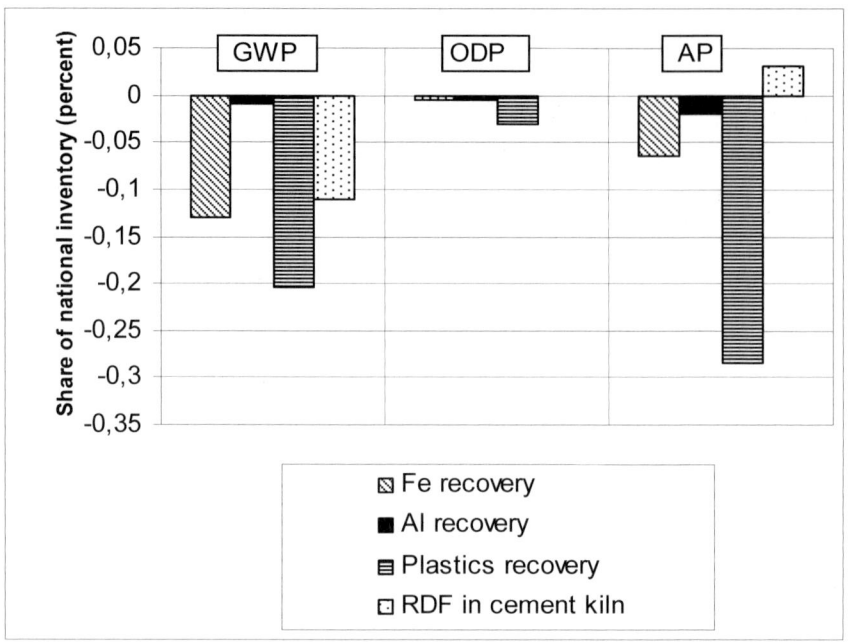

Figure 12.8 Results of a complex LCA of MBP recovery options (Soyez, 2001)

According to figure 12.8 in most of the criteria environmental benefits (i.e. negative values in the figure) result even if one considers the additional impacts for the separation and recovery processes.

Climate effects of all options are beneficial. Best results will be by material recycling of plastics provided a high value end use of the recycled plastics is guaranteed. The benefits of ferrous metals recovery are as high as those of using the RDF fraction in a cement kiln, and the power production from biogas. The improvement potential of the aluminum recovery is of minor relevance compared to the other recovery options.

The benefits in the impact categories result from the substitution of primary materials and energy resources which otherwise would have to be produced and used with the corresponding environmental impacts. But the recycling oriented residual waste treatment strategy requires a market which is open for the recovered resources and a material management which ensures that the recovered materials comply with the requirements of the recycling industry. It is, for example, very important to use the waste heat of RDF incineration and the biogas burning processes, otherwise the ecological benefit is very small.

Climate Effects of Mechanical-biological Waste Pre-treatment   167

12.6.2.2   Optimising the climate impacts of the MBP technology

As was shown in figure 12.6, MPB technology in addition to the recycling material including the RDF fraction leads to a product which after biological treatment is to be deposited. For suitable information on climate effects of MPB as a whole, the effect of the deposited material on the landfill has to be taken into account. A better pre-treatment comes up with lower organic residues and thus with lower landfill gas emissions. On the other hand, longer treatment requires higher energy input, e.g. for aeration, which as a consequence leads to enhanced greenhouse gas emissions in this process step. This is a typical technological compromise situation (see chapter 8) for which the best solution has to be found.

Figure 12.9 displays GHG emissions by energy consumption as well as landfill methane emission with the process duration. The sum of both curves shows a minimum which defines the best decomposition duration. Obviously it is not useful to stabilise the waste to the maximum degree as the overall result gets worse with longer process duration due to a higher total energy input. As dependent on the methane oxidation capacity of the landfill cover the minimum GHG value and therefore best process duration is between 16 and 28 hours. For 95 percent methane oxidation in the landfill layer 18 weeks is the best option.

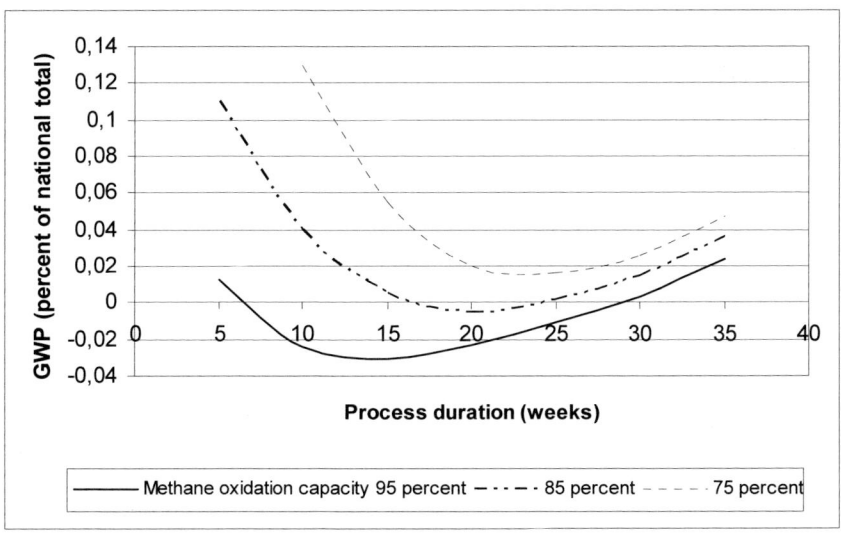

Figure 12.9   GWP as dependent on process duration (Soyez, 2001)

12.6.2.3 Climate effects of MBP waste gas treatment

Waste raw gas from the biological treatment process of a MBP facility contains a broad variety of climate relevant substances i) from constituents of the waste as well as ii) from substances which are caused by the biological process. The total emission potential of organic substances (TOC, total organic content) in the raw gas is about 1 kg per ton of waste, whereas nitrogen compounds amount about 0.1 kg/t wet waste. To comply with the law emissions must be reduced by gas cleaning. There are several techniques to reduce the emissions such as biofilters and Regenerative Thermal Oxidation (RTO) which both are broadly applied in state-of-the-art MBPs. Only is the RTO technology able to reduce the residual organic load in waste gas to the level of the allocation values (see table 12.12).

In the RTO technology the waste gas is oxidized by co-combustion of natural gas or liquid gas. As a result, most of the organic components are destroyed. The residual specific load of organics in the waste gas is far less than 55 g per ton of waste product. However the combustion of the gas is combined with emissions of CO_2 as well as with methane emissions in the gas processing chain, which have to be considered in the total budget. Incineration in the RTO technology furthermore results is N_2O emissions.

Evaluating the net effect under climate respect the benefits as well as the drawbacks have to be considered. An example is given in table 12.13 for two technological options applied in two RTO facilities. In one facility natural gas is used, in the other case, LFG from a nearby landfill.

Table 12.13 Comparison of GHG effects of RTO variants (Zeschmar-Lahl, 2000)

RTO process step	GHG effect (kg CO_2-eq./t)	
	Facility 1	Facility 2
Natural gas pre-processing	0.21	0.0015
CO_2 from natural gas incineration	6.17	0.44
Power consumption	3.4	3.4
CO_2 from mineralization of organics	0.62	0.62
Effects of N_2O deliberation	1.9	1.9
Mineralization of HFCs	-8.67	-8.67
Oxidation of methane	-4.45	-4.45
Benefits of landfill gas oxidation	0	-69.2
Total	**-0.82**	**-75.95**

By application of landfill gas instead of natural gas huge benefits result for the process. About 76 kg CO_2-eq. are saved per ton of waste treated compared to facility 1, where the net benefit is less than one kg. This is close to the break-even-point where no positive climate effect is achieved by application of the RTO process. Such a case may occur, for example, if the amount of HFCs is by about 10 percent less, e.g. as a long term result of phasing-out of these substances under the provisions of the *Copenhagen Amendments to the Montreal Protocol* (see chaper 10).

As a message from this it can be drawn that a technology able to reduce emissions of certain components must be totally budgeted to be sure that a net benefit can be achieved even under changing process conditions in the long term.

12.7 WARM – a Tool for GHG Evaluation of Waste Management Strategies

WARM (**WA**ste **R**eduction **M**odel) was created by the U.S. Environmental Protection Agency (EPA) to support solid waste managers and organizations in planning waste management strategies under climate aspect. It is available in a web based calculator format and as a Microsoft EXCEL spreadsheet (EPA, 2005a).

WARM calculates GHG emissions for waste management practices, including source reduction, recycling, combustion, composting and deposition. In every calculation case a baseline and an alternative option are compared. The GHG emission factors were calculated following the LCA methodology (see chapter 8). A wide range of materials is considered (see table 12.14). For an explanation why recycling some materials reduces GHG emissions more than source reduction (e.g. aluminium) see EPA (2008b).

Table 12.14   GHG emission data of waste components used in the WARM model (EPA, 2005a)

Material	GHG emissions of materials (t CO_2-eq./t)				
	source reduced	recycled	deposited	combusted	composted
Aluminum cans	-8.97	-14.93	0.04	0.06	
Steel cans	-3.21	-1.79	0.04	-1.53	
Copper wire	-7.55	-5.08	0.04	0.06	
Glass	-0.58	-0.28	0.04	0.05	
HDPE	-1.81	-1.41	0.04	0.90	

Material	GHG emissions of materials (t CO$_2$-eq./t)				
	source reduced	recycled	deposited	combusted	composted
LDPE	-2.29	-1.71	0.04	0.90	
PET	-2.12	-1.55	0.04	1.07	
Corrugated cardboard	-2.63	-2.74	0.59	-0.66	
Magazines/third-class mail	-4.30	-2.70	-0.23	-0.48	
Newspaper	-4.06	-3.49	-0.80	-0.75	
Office paper	-3.64	-2.48	2.27	-0.63	
Phonebooks	-5.23	-3.34	-0.80	-0.75	
Textbooks	-4.82	-2.74	2.27	-0.63	
Dimensional lumber	-2.02	-2.45	-0.39	-0.79	
Medium density fiberboard (MDF)	-2.23	-2.47	-0.39	-0.79	
Food scraps			0.84	-0.18	-0.20
Yard trimmings			-0.15	-0.22	-0.20
Grass			0.03	-0.22	-0.20
Leaves			-0.10	-0.22	-0.20
Branches			-0.39	-0.22	-0.20
Mixed paper, broad		-3.17	0.52	-0.66	
Mixed paper, residential		-3.17	0.42	-0.66	
Mixed paper, office		-3.06	0.64	-0.60	
Mixed metals		-7.27	0.04	-0.47	
Mixed plastics		-1.51	0.04	0.97	
Mixed recyclables		-2.87	0.28	-0.62	
Mixed organics			0.33	-0.20	-0.20
Mixed MSW			0.58	-0.13	
Carpet	-4.10	-7.36	0.04	0.37	
Personal computers	-58.07	-2.46	0.04	-0.20	
Clay bricks	-0.29		0.04		
Aggregate		-0.01	0.04		
Fly ash		-0.87	0.04		

To use the model it is necessary to define the waste management practices to be compared and to gather the waste management data, such as type and amount of waste components at the existing waste management practice as well as in prospective alternative scenarios. For all technological elements of waste management processes WARM proposes certain values (see table 12.15) which may be changed if necessary to adapt to a specific situation.

Table 12.15   WARM standard technological items

Waste management practice	Program standard	Possible choices by user
Benefits of source reduction	Current mix of virgin and recycled inputs (see figure 12.1)	100 percent unhandled material (represents an upper limit of possible effects)
Landfill gas recovery	U.S. national average based on the emission proportions of landfills with landfill gas control in 2000	Landfill gas recovery No landfill gas
		Recovery of energy Flare only
	Landfill gas efficiency after national average: 75 percent	Current or predicted efficiency
Waste transportation distances	Estimated distances from the curb to the landfill or waste treatment facility, as combustion, recycling or composting: 20 miles	Actual distance

WARM calculation example

As a calculation example a waste is considered which consists of aluminum and steel cans, glass, plastics (HTPE, PET), corrugated cardboard, as well as three types of organic matter, such as yard trimmings, grass, and leaves. The composition of the waste is given in table 12.17 for six scenarios. The total mass processed is 800 t.

The greenhouse gas effects were calculated under the following assumptions:

- The waste is deposited in a landfill which is equipped with a landfill gas recovery system for electricity production. Landfill gas collection efficiency is 75 percent.
- Greenhouse gas emissions of deposition, recycling, combustion, composting, and source reduction are calculated on the basis of table 12.14 (EPA, 2005a).
- Benefits of recycling are calculated as compared with the current mix of recycled and virgin matter in manufacturing (see figure 12.1).

The following scenarios (see table 12.16, 2nd column) were studied under the specification of a waste composition based on table 12.17:

Table 12.16 Description of scenarios and resulting GHG reduction

Scenario	Description	Resulting GHG reduction (t CO_2-eq.)
0	Reference scenario: whole waste deposited	0
1	Recycling of 50 percent of the waste components, without organic green matter	1,128
2	Same as 1, but total composting of the organic matter	1,165
3	Same as 2, but 50 percent of the plastics and the corrugated cardboard combusted	1,134
4	Same as 3, but no recycling, 100 percent combusted	783
5	Same as 2, but no combustion, 100 percent recycling	1,484
6	Same as 2, but 25 percent of aluminum and steel cans as well as glass deposited, and 25 percent source reduced	1,764

Table 12.17 Waste composition and mass balances of the scenarios (in t)

Waste material	Waste mass	Scenario						
		0	1	2	3	4	5	6
			Recycling	Composting	Combustion	Combustion	Recycling	Source reduced
Aluminium cans	100		50					25
Steel cans	60		30					15
Glass	40		20					10
HDPE	100		50		50	100	100	
PET	100		50		50	100	100	
Corrugated cardboard	100		50		50	100	100	
Yard trimmings	100			100				
Grass	100			100				
Leaves	100			100				

The results are given in figure 12.10.

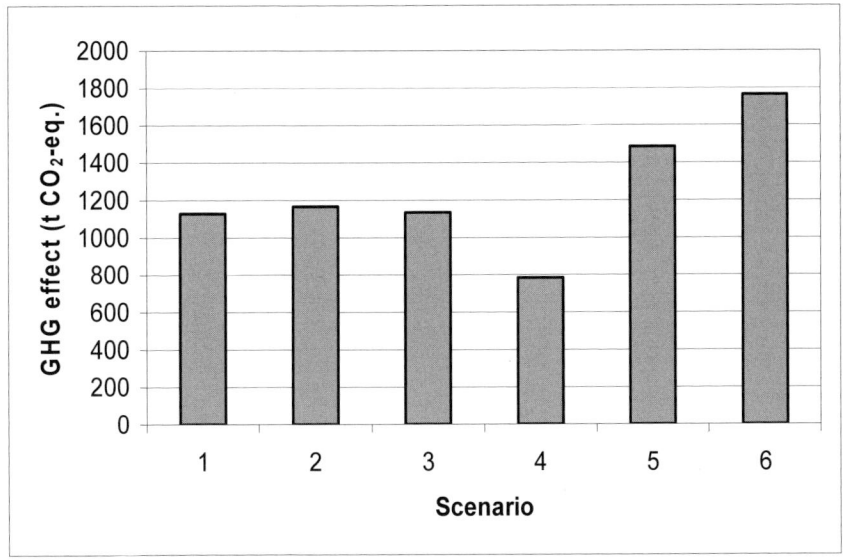

Figure 12.10 GHG reduction by waste management scenarios

Figure 12.10 indicates that each waste scenario other than deposition on the landfill (reference scenario) considerably improves the greenhouse gas balance.

The recycling of 50 percent of metals, glass, and plastics results in 1,128 t CO_2-eq. reduction (Scenario 1). In this scenario, organics are deposited. If organics are fully composted (Scenario 2), another small improvement in the CO_2 budget will occur (1,165). If 50 percent of the plastics are combusted instead of deposited (Scenario 3), nearly no effect will result as compared with scenario 1. Obviously, in this case, the GHG emissions resulting from landfill or from combustion are nearly the same.

However, if a fraction of 50 percent of plastics is combusted instead of its recycling (Scenario 4), the result will be much worse (783), since GHG reductions caused by recycling no longer exist. The considerable contribution of recycling to GHG reduction would become even more evident if a 100 percent recycling is considered (Scenario 5). In this case, a high reduction in GHG emissions occurs (1,484). An even more positive effect (1,764) would be achieved if a portion of the metal and glass fractions is source reduced (Scenario 6). In the case considered, 50 percent are recycled, but 25 percent are deposited, and the other 25 percent source reduced.

13 Energy Related Climate Impacts

The world's current annual primary energy use is about 470 EJ whereof 86.8 percent are fossil and 13.2 percent are renewable in origin (see table 13.1). The renewables comprise an about 83 percent share of biomass.

Table 13.1 Current global primary energy sources (IPCC, 2007c)

Energy source	Fossil				Renewable				
	Oil	Coal	Gas	Nuclear	Biomass	Hydro	Geothermal	Solar	Wind
EJ/a	165	116.5	95.3	29.9	49.6	10.1	0.9	0.8	0.7
Percent	35.2	24.9	20.3	6.4	10.6	2.2	0.2	0.1	0.1

Of final consumer energy 45 percent is used for low temperature heat such as cooking, water and space heating, and drying; 10 percent for industrial process heat; 15 percent for electric motors, lighting, and electronics; as well as 30 percent for transport. Around one half of the total is consumed by the developed countries which comprise one billion people. Another billion in the poorest countries consume less than 4 percent. In all regions of the world as well as all sectors of energy use, demand has grown in recent years. Only an increase by two percent was recorded between 2000 and 2002. Two-thirds more energy need is anticipated for 2030 compared to the year 2004 if no action to change energy use patterns is taken (IPCC, 2007c).

There are two consequences seen: i) Rising demand implies a resource problem since current energy sources are mostly of fossil origin and therefore of limited availability. However ii) the impacts of greenhouse gas emissions from burning of fossil energy sources are of even larger importance since – as was shown in figure 9.1 – energy related GHG emissions are most important contributers to world's total GHG balance. Fossil energy use is therefore the main reason for climate change. Improved energy efficiency (i.e. energy use per GNP), the switch to other energy sources especially of renewable origin, as well as changed energy use patterns are urgently necessary and are high ranking in the political agenda as well as in industry and the other production spheres.

The following chapters specify the GHG emissions from fossil fuel, its extraction and its use, and the implications of a shift towards renewable energy. For basic considerations on the potential and the status of renewable energy see also chapters 6.4.1. and 6.4.2. Examples of improved energy efficiency measures are discussed in all other chapters.

13.1 GHG emission overview

Direct GHG emissions from fossil energy related activities are mainly caused by the direct greenhouse gases CO_2, CH_4 and N_2O. The majority come from fossil fuel combustion with CO_2 as the primary gas emitted in an amount of about 27,100 Mio t CO_2-eq. per year on a global scale, of which 39.1, 23.2 and 37.6 percent are from coal, oil and gas, respectively (IPCC, 2007c). The total amount of CO_2 which was emitted during the last 150 years is estimated to be about 1,100,000 Mio t!

The contribution of the national economies is very diverse in total, structure, and in per capita numbers, which is documented by the following selected data:

- The United States contribute by about 22 percent to the global energy related CO_2 emissions. Energy related activities accounted for over 85 percent of the country's total emissions (in 2005), where CO_2 emissions account for nearly 82 percent compared to only 4 percent of non-CO_2-emissions. 98, 38, and 11 percent of the nation's CO_2, CH_4 and N_2O emissions, respectively, are included in this number. U.S. power plant GHG emissions total 2,650 Mio t CO_2-eq. (EPA, 2007b).

- Africa's fossil-fuel CO_2 emissions are low in both absolute and per capita terms. They were below 900 Mio ton of CO_2-eq. in 2002. Per capita emissions are only about 5 percent of the comparable value for North America. Fuels account for 15.6 percent. The emissions are mostly due to the activities of only a few countries, amongst them South Africa which accounts for 40 percent of the continental total. Another 44 percent share is from Egypt, Algeria, Nigeria, Libya, and Morocco, combined. Power in South Africa largely depends on coal burning.

- In Asia from 1971 to 1995 an eightfold increase in coal burning for electricity generation in the industrialising countries occured. The trend continues, and India and China will be responsible for 75 percent of the increased global coal consumption. By 2025, nearly 60 percent of all coal will be burned in Asia-Pacific. As a consequence, the region's CO_2 emissions are predicted to double (WWF, 2006). In Thailand the total consumption of fossil fuels is about 232

Mio t CO_2-eq. annually (in 2002). The consumption more than doubles every 10 years.

In addition to emissions which are directly related to the use of fuels, greenhouse gases are released by energy related activities such as mining and production, transmission, storage, and distribution. These emissions mostly originate from fugitive methane from natural gas and petroleum systems, and coal mining. During these processes, CO_2 as a direct greenhouse gas, as well as the indirect greenhouse gases such as CO, NO_x, NMVOCs, and SO_2 are emitted, but in smaller quantities.

Emissions can be allocated to end use sectors, such as transportation, industry, residential and commercial. Figure 13.1 displays GHG emissions by these sectors with specific contributions of both combustion and electricity related emissions. The industrial end use sector includes activities such as manufacturing, construction, mining, and agriculture. Manufacturing is the largest consumer of energy. Six industries, comprising petroleum refineries, chemicals, primary metals, paper, food, and non-metallic mineral products, represent the majority of energy use. In the United States, this sector accounts for 28 percent of CO_2 emissions from total energy use. A share of one-half results from direct consumption of fossil fuels for steam and process heat, whereas the other half is used for electricity for motors, electrical furnaces, and lighting – see again figure 13.1, second column.

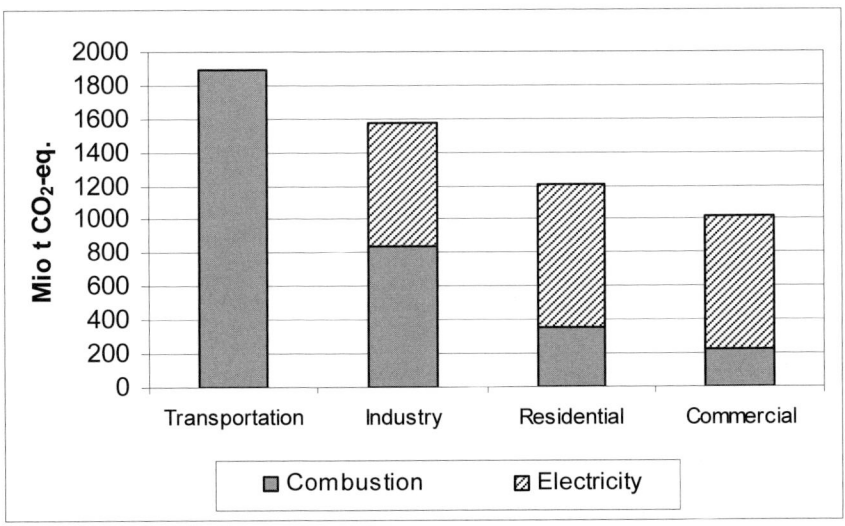

Figure 13.1   **GHG emissions from fossil fuel combustion by end use sectors in the USA (in 2005; EPA, 2007b)**

13.2 GHG Emissions by Extraction of Fossil Fuels

Extraction technologies refer to mining of coal and extraction of gas and oil. Greenhouse effects predominantly occur through methane emissions.

13.2.1 Methane emissions by oil and natural gas extraction

Production, processing, transmission, and distribution of oil and natural gas are the second largest anthropogenic methane source globally. About 88 billion m^3 are released annually. In Russia and Ukraine, in 2000 an equivalent of 69.1 and 16.4 Mio t CO_2, respectively, was set free. In the U.S. oil industry by petroleum systems in 2005 methane emissions were about 28 Mio t CO_2-eq. (EPA, 2007b).

In **petroleum systems** methane emissions are primarily associated with crude oil production, transportation, and refining operations. Methane is released as fugitive emissions, vented emissions, from operational upsets, and from fuel combustion. Production field operations account for over 97 percent, from which vented emissions are 90 percent. Most dominant are offshore oil platforms, field storage tanks, and natural gas powered pneumatic devices.

Methane losses from **natural gas systems** account for 15 percent of total worldwide methane emissions. Emissions vary greatly from facility to facility and are largely dependent on operation and maintenance procedures and equipment conditions. Primary contributors are normal operations, maintenance, and system upsets. In field production which is the initial stage of the natural gas system, one-third of the total emissions are released. Fugitive and the emissions from pneumatic devices are major sources. During the processing, where a pipeline quality gas is produced, further 12 percent of the emissions occur, mostly fugitive. Methane emissions from transmission and storage account for another third of the emissions. The distribution system, through which the natural gas is transported to the end user, also accounts for one-third of the emissions.

Considerable reduction is possible through upgrading of technologies or equipment and by improved management and operational procedures. Examples are low-emission regulator valves, operational procedures to reduce venting, leak detection, and more precise measurement.

An example of possible effects of such measures is an emission detection program at *Cherkasytransgas,* one of six Ukrainian gas system subsidiaries. After running measurement studies it was determined, that 3 Mio m^3/a natural gas were leaking from only two compressor station sites. After repairing, leakage of two-thirds of this amount was prevented. The programme continues to permanently monitor and improve all compressor stations (Mandra, 2004).

13.2.2 Methane emissions from coal mines

Coal mining releases methane by underground and surface mining as well as by coal handling (post mining) activities. Since methane can create an explosive mixture with oxygen it must be removed for safety reasons by large-scale ventilation systems. They move massive quantities of air through the mines. Only the gas contains methane in very low concentration. Underground mining is the most important source of fugitive mine methane by far with about 70 percent of all coal mining related emissions. Methane is also produced from degasification systems (gas drainage systems) that employ wells to recover methane.

The background of methane from coal is as follows: Methane in coal was produced when vegetation was converted during the coalification process in earlier periods of Earth's history. It was stored under high pressure within coal seams and surrounding rock. It is liberated when mined. The quantity of the methane emitted depends on the carbon content in the coal (coal rank) and the coal depth. Coals such as anthracite have the highest carbon content and release most methane. Lignite is lowest in methane deliberation. Specific methane emissions range from about 0.7 to over 310 m^3 per ton, e.g. in the U.S. mining industry.

Globally, coal mine methane accounts for 8 percent of total methane emissions resulting from human activities. In 2000, worldwide emissions totaled 380 Mio t CO_2-eq. By 2020, the world's coal mines are expected to produce annual emissions of 450 Mio t CO_2-eq./a.

More than 90 percent of the world's methane from coal mines is from only 11 countries. Amongst them China (40 percent) and USA (14 percent) as the biggest coal producers, together are emitting more than 50 percent of the world's total. Other significant emittents include Russia and Ukraine (7 percent each), Northern Korea (6 percent) and Australia (4 percent). India, Germany, Kazakhstan, South Africa, and the United Kingdom contribute 2 to 3 percent each. In some countries, such as Mexico and Vietnam, emissions of methane from mines are low on a national level, but some of their coal mines produce substantial volumes (EPA, 2007e), and are worthy of note therefore.

For decades methane in coal mines was considered a burden and a costly safety hazard. But recent projects have proven that it can provide many benefits to the mine and the environment. A variety of uses is possible. The optimal use at a given location is dependent on the quality of the methane containing gas, the availability of end-use options and the project economy. Possible uses include natural gas pipeline injection, electric power production, district heating, co-firing in boilers, mine heating, coal drying, vehicle fuel, flaring, and manufacturing or industrial uses such as feedstock for carbon black, methanol, and dimethyl ether production. Low concentration methane can be oxidized and the resulting thermal energy can be used to produce heat, electricity, and refrigeration.

The following examples will give an overview of the situation in the most important countries and initiatives to reduce mining methane emissions.

- In China it is estimated that there are more than 26,000 coal mines, mostly (90 percent) underground, which produce about 1.4 billion t of coal (2003). 50 percent of the large mines are emitting gas; they account for nearly 50 percent of coal production and about 86 percent of coal mine methane emissions. By 2004 over 200 mines were equipped with methane drainage systems. Over 1.5 billion m^3 of methane had been recovered with approximately 40 percent of drained gas utilized. The gas is used for town gas, power generation, industrial applications, and vehicle fuel (EPA, 2007f).

- An example of a successful Chinese project is the Jincheng Anthracite Mining Group which produces anthracite coal at several highly gassing mines. They started methane capturing in 1995. The gas was used to fuel a 1.6 MW power station. Capacity was increased to 4 MW in 2002. The effect on GHG emissions is estimated to be about 40,000 t CO_2-eq. per year. As a next step a 120 MW power station using the latest technology will be established, which by 2008 will serve the total power needs of the mine as well as 90,000 households and commercial and industrial consumers in the area. The project will save emissions of 7 Mio t CO_2-eq. annually.

- U.S. mining industry accounts for about 10 percent of man-made methane emissions. In 2003 4.9 billion m^3 were liberated from which one-quarter was recovered and used. GHG effects from U.S. coal mining methane declined by 170 Mio t CO_2-eq. from 1990 to 2005. About 15 projects are in operation at active mines, and more than 20 in abandoned mines. The gas is used for mine heating, pipeline injection and power generation. Gas sales generate about US$ 90 Mio in revenues annually.

- For further improvement of the situation on an international level a non-profit organization "Methane to Markets initiative" was established in recent years. All important coal producing countries and regions are members, including USA, EU, and China. It aims to improve awareness of emission reduction opportunities and the value of the recovered methane as well as to implement projects worldwide. A data submission form for project feasibility studies (EPA, 2007e) as well as a summary of methane technologies for mitigation and utilisation was prepared to support decision making processes (EPA, 2007e).

13.3 Climate Effects of Fossil Fuel Use

Fossil fuel is used in every sector of the economy, amongst which power production, industry related fuel use, and transportation are the most important and therefore will be discussed in the following chapters.

13.3.1 Climate effects of fuel use in power production

Power is produced from a variety of fuels using diverse technologies. The concrete situation in a country's energy sector depends on factors such as the available energy basis and established industrial facilities, but also public opinion or legislation.

As an example of the energy basis the current fuel mix in the U.S., German, and South African power industry is displayed in figure 13.2. Coal dominates the electricity production in each of the countries however South Africa with a share of more than 90 percent depends largely on only this type of fossil sources.

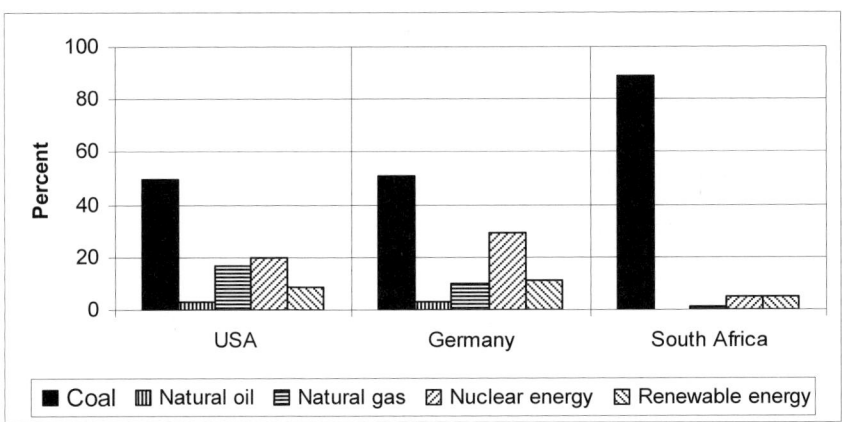

Figure 13.2   **Power production energy mix in selected countries**

In accordance with the composition of the fuel mix the mean emission equivalent, i.e. GHG emissions per energy unit, varies in different countries. E.g. in Germany the value is 514 g CO_2-eq./kWh compared to South Africa's 900 g, which is nearly twice (Soyez, 2008).

The reason is that specific emissions to a large share depend on the type of fuel used. In the case of nuclear power and renewable resources considerably lower

GHG amounts per unit of electricity produced are emitted as compared with fossil fuels (see chapter 13.4). However also emit fossil fuel types different amounts of GHGs due to their carbon content and the fraction of carbon that is oxidized during combustion. Theoretical emission values are given in figure 13.3. The amount of carbon not oxidised due to inefficiencies during the combustion process is less than three percent in modern coal power stations, one percent for petroleum and half a percent for natural gas.

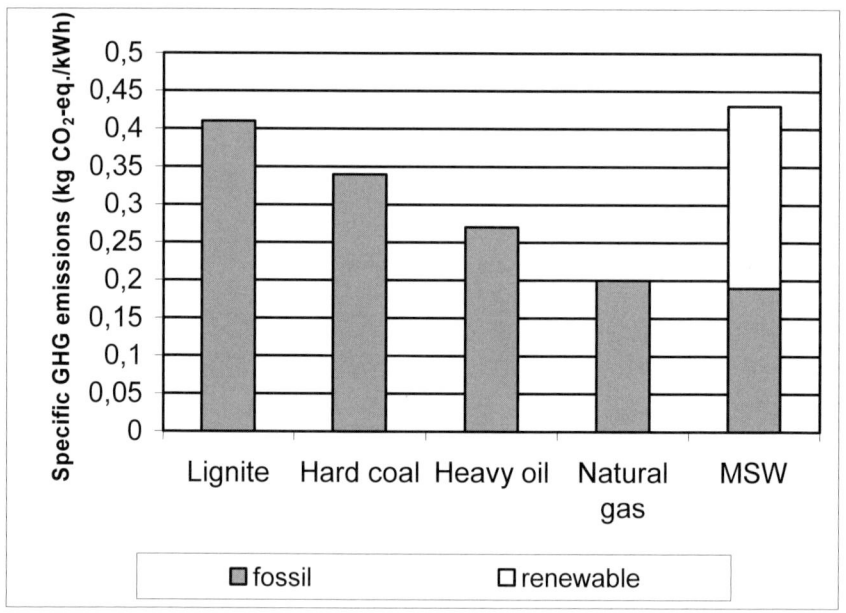

Figure 13.3   **Specific CO_2 emissions from fossil fuel use**

Besides direct CO_2 emissions also CH_4 and N_2O must be considered. Methane emissions are 1.5 kg/TJ for solid, 3.5 kg/TJ for liquid, and 0.3 kg/TJ for gaseous fuels. N_2O emissions after several technologies are given in table 13.2.

Table 13.2   **Technological N_2O emission factors from large combustion plants (UBA, 2007)**

Fuel	Combustion technology	N_2O emission factor (kg/TJ)	GHG effect (t CO_2-eq./TJ)
Hard coal	Fluidized bed	20	5.96
Hard coal	Other combustion methods	4	1.19
Lignite	Fludized bed	8	2.38

Fuel	Combustion technology	N$_2$O emission factor (kg/TJ)	GHG effect (t CO$_2$-eq./TJ)
Lignite	Dry-dust combustion	3.2	0.95
Lignite	Other combustion methods	3.5	1.04
Liquid fuels		1	0.3
Gaseous fuels		0.5	0.15

An improvement of the GHG emission situation would therefore be possible by a change of the energy basis. For such a shift the term "fuel switching" is common. Actually best results can be achieved by switching from fossil to renewable resources which have very slow GHG emissions (for values see table 13.4). However also in the case of a change from coal and oil to natural gas the emissions are considerably reduced; e.g. from lignite to gas cut in half. The "fossil fuel switching" is therefore a very important measure to reduce GHG emissions. This option currently is seen as the most effective possible short term measure to slow down the climate change rate.

As another option of fuel switching the co-firing of MSW seems promising even though the specific GHG emissions are highest compared to lignite (see figure 13.3). The reason is that only a part is of fossil origin which has to be considered GHG relevant. The other portion is of renewable origin and thus climate neutral (see chapter 13.5). The fossil based GHG emissions are similar to natural gas, since the fossil portion of waste mostly consists of plastics which have a chemical composition comparable with natural gas. Use of waste is therefore sometimes seen as an attractive option for low carbon power production.

In addition to fuel switching upgrading of technology and equipment as well as improved management and operational procedures imply a large potential for emission reduction in the energy sector. The following example illustrates this fact in the case of lignite fired power stations: For the most dirty 30 plants worldwide values up to 1,350 g CO$_2$-eq./kWh were measured after full balancing of all processes of electricity production (WWF, 2006). This is 3.5 times the value given in figure 13.3. One concrete measure to improve the efficiency is the use of waste heat from thermal electricity generation plants in industry or for buildings and district heating.

Highest effects can be achieved by combined measures. As an example, in the case of the power industry in the EU during 1990-2004, only emissions increased by about 6 percent during 1990-2004 whereas 40 percent more electricity was generated. This means that emissions were no longer in line with energy generation. The improved relationship between the actual emission reduction and the increased power generation can be explained by i) fuel switching from coal and lignite to natural gas which resulted in 18 percent reduction in specific GHG emissions, and ii) 12 percent reduction by improved thermal efficiency of electricity and heat production (EC, 2007).

In the case of one German supplier a specific value of 12 g CO_2-eq./kWh compared to the German average of 514 g CO_2-eq./kWh was established by use of renewables and combined heat and power production (cogeneration) where waste heat was used for buildings and district heating (Schönau, 2007).

13.3.2 Climate effects of transportation

In terms of emission levels, aircraft, road, and railway transportation are very important climate impact sources. However there are considerable differences in the climate impacts of the types of transportation: One ton of CO_2 emissions allows for a travel distance of 3,000; 7,000; and 17,000 km by aircraft, car, and train transportation, respectively. Train transportation of goods is characterised by the lowest climate impacts with 29 g CO_2-eq. per ton and km compared to 158 for road haulage and 31 for shipping (under German conditions; Allianz, 2003). Moreover, in the case of aircraft transportation, not only CO_2 emissions count. There is an extra climate burden by water vapour and pollutants at great heights.

Some details of road and aircraft transportation are discussed below.

13.3.2.1 Road transportation

Total actual emissions from passenger cars are affected by the number of cars in use and the average annual mileage, the average specific fuel consumption, driving styles and auxiliary installations.

Worldwide, 600 million cars are driven, 75 percent of which in developed countries. Transport CO_2 emissions in the EU grew by 32 percent between 1990 and 2004. The share of transport in CO_2 emissions was 21 percent by 1990; by 2004 this had grown to 28 percent. Emissions from passenger cars and vans account for approximately half of this (T&E, 2006). In Germany road emissions contributed about 15 percent to the total national emissions which was an amount of 150 Mio t CO_2-eq. in 2005 (UBA, 2007). In California 60 percent of the total GHG emissions are caused by road traffic (Sand, 2005).

Emissions are caused by direct CO_2 emissions from gasoline and diesel fuel as well as greenhouse gases other than CO_2, including CH_4, N_2O, and the indirect greenhouse gases NO_x, CO, NMVOCs, NH_3, and SO_2.

CO_2 emissions. The emission value for CO_2 can be calculated from the amount of fuel used applying a factor of 3.175 kg CO_2 per kg fuel (or 2.4 kg per liter fuel at a specific weight of gasoline of 0.742 kg/l). For gasoline driven cars and a current mean fuel consumption of 8.5 l, there specific emissions of 200 g CO_2 per km result. A value of 50 percent less is possible for new cars taking into accout a consumption of 3 to 4 l because of advanced standards of engine development. To reach such a goal there were several measures planned by policy as well as by the car producers:

In 1996, the EU approved a *"Community Strategy to reduce CO_2 emissions from passenger cars"*. The objective was to reduce the average CO_2 emissions of new passenger cars in the EU to 120 g/km by 2005, or 2010 at the latest. This target represents a 35 percent reduction over 1995 levels and corresponds to a fuel consumption figure of 5 and 4.5 liters per 100 km for petrol and diesel cars, respectively, as measured on the official European Driving Cycle. This objective was to be reached through technical measures, consumer information, and fiscal initiatives. In addition to reduced specific fuel consumption (liter per km) the supplementing of fossil fuels by biofuels is seen as the most favourable measure to reduce emissions in traffic-related CO_2 emissions (see chapter 13.5). A recent EC initiative which is under controverse discussion calls for 110 g per km including a 10 g/km credit of biofuels (EC, 2008).

In 1998 the European Automobile Manufacturers Association (ACEA) committed the EU to reducing average CO_2 emissions from their new car sales in the EU to 140 g per km, by 2008. This is equivalent to 6.0 and 5.3 liters per 100 km for petrol cars and diesel cars, respectively, and means a reduction of 25 percent over 1995 levels. A similar commitment was made by the Japan (JAMA) and the Korean (KAMA) Automobile Manufacturers Associations for their EU sales to achieve this figure in 2009. These aims were not yet achieved.

The specific fuel consumption measured according to Directive 93/116/EC does not include fuel used for powering electric equipment such as headlights, electrically warmed seats, or air conditioners, for which an extra fuel consumption between 10 and 15 percent (Kageson, 2005) must be taken into account to establish a real figure of consumption.

Other emissions. Emissions of CH_4, N_2O, NO_x, CO, NH_3, NMVOC, and SO_2 decreased sharply in recent years due to improved fuel quality, use of catalytic converters and engine improvements resulting from more stringent emissions laws. N_2O and NO_x emissions are closely related to fuel characteristics, air/fuel mixes, combustion temperatures, as well as usage of pollutant control equipment.

N_2O emissions result primarily from incomplete reduction of NO to N_2 in 3-way catalytic converters, which aim to control NO_x, CO, and hydrocarbon emissions. Older values for N_2O and NH_3 are in the range of 10 to 43 mg/km and 10 to 30 for gasoline after EURO-4 and EURO-1, respectively. Values for diesel fuel are much lower, 8 and 1 mg/km, respectively. CO emissions are affected by combustion efficiency and post-combustion emission control equipment. They are highest when air-fuel mixes have less oxygen than required for 100 percent combustion, e.g. in idle, low speed, and cold start conditions. Methane and NMVOC emissions depend on the CH_4 content of the motor fuel, the amount of uncombusted hydrocarbons, and the efficiency of post-combustion converters.

The value of all emissions strongly depends on individual driving behaviour. With reference to CO_2, a lower consumption is possible by speed reduction, e.g. to a mean value of not more than 100 km per hour on longer distances. Furthermore, short distance driving of only a few kilometers should be avoided (see chapter 14).

13.3.2.2 Aircraft transportation

Air transport emissions cause 6 percent of the GHG emissions worldwide or 12 percent of the world's transport sector. Five percent growth in total aircraft transportation (in km) is estimated per year.

Air transportation affects climate for several reasons: Carbon dioxide emissions from kerosine depend on fuel consumption. As a positive result, specific emissions were reduced by about 20 percent in the last decade. However, total emissions have been increasing continually as a result of increasing air transport. CO_2 emissions are expected to double in the years from 1990 to 2015 and to multiply threefold by 2050.

In addition to CO_2, water vapor and nitrogen oxides as well as, secondarily, hydrocarbons such as NMVOCs, particulates, CO, nitrous oxide, and sulphur dioxide are of considerable significance. Emission factors are given in table 13.3.

Table 13.3   Emission factors for civil aircraft transports (UBA, 2007)

Component	Specific emission	
	g/kg kerosin	kg/TJ
CO_2	3,299.7	74,265
NO_2	17.4	390
CO	17.4	390
NMVOC	2.60	59
SO_2	0.2	4.7
N_2O	0.1	1.5
CH_4	0.04	1

A third factor must be considered which is that the impact effects on climate from aircraft are different from those caused by ground transportation. Focusing on CO_2 emissions alone is not sufficient. Due to the emission of water vapour condensation trails and cirrus clouds are formed in the higher atmosphere, which influence regional climate (see chapter 2). This fact is especially important if flight height is more than 10,000 m.

Summing up the climate impact which is expressed by a so-called *Radiative Forcing Index (RFI)* is estimated to be two- to fourfold of CO_2 emissions. Considering this value which is under current scientific discussion, the contribution of aircraft transportation is about 10 percent of the total climate impacts. Adverse effects such as cooling by the formation of clouds from SO_2 aerosols seem possible but will not compensate for the warming effects.

Actual improvements seem possible mostly through further reduced fuel consumption. In the case of Lufthansa a current value of 4.4 l per 100 km and person will be reduced to 3.0 l in the coming years by new aircraft. Furthermore, improved organizational measures such as European Single Sky which will become a reality in 2020 or direct flight instead of air corridors which together are expected to reduce GHG emissions by 8 to 12 percent are proposed (Noack, 2007). However other environmental as well as health issues have to be taken into account.

A shift to less GHG emitting fuels is a future option which is already under discussion, e.g. by EADS for its Airbus 380 version for which an alternative liquid quality fuel made from natural gas was recently tested. Current problems in case of biofuels are lower energy per mass ratio of fuel, limited availability in sufficient commercial quantities, and safety constraints. The company announced to study viable 2nd generation biofuels when they become available (EADS, 2008). In Brazil a 100 percent bioethanol fuel (E100) has been marketed for use in small aircraft since 2004 (Szwarc, 2004).

13.4 Climate Effects of Other Non-biomass Energy Sources

Energy sources other than coal, oil and gas comprise nuclear power and renewable energy sources such as wind, hydropower, geothermal energy, wave energy, solar radiation and biomass. Their shares of primary energy sources were given in table 13.1. Amongst them biomass related energy is discussed in more detail in chapter 13.5. The following refers mainly to the non-fossil-fuel but non-biomass energies.

Specific carbon dioxide emission factors of such energy sources are theoretically 0 g CO_2-eq./kWh, for they do not result in direct emissions. However, indirect emissions from life cycle activities such as construction, material input, and maintenance may not be negligible in most cases. A climate-related advantage of non-fossil fuel based energy sources therefore cannot be postulated in general. This is demonstrated by some numbers of GHG emissions given in table 13.4.

Table 13.4 CO_2 emission factors for non-fossil energy sources

Energy source	Emission (g CO_2-eq./kWh)		
	(Fritsche, 2007)	1999 (Marheineke, 1999)	Future perspective (Marheineke, 1999)
Nuclear power	8-120	18-20	18-19
Wind power	23-24	17-28	
Photovoltaic	101	193-336	129-131
Hydropower	40	10-17	

From this the following conclusions can be drawn:

- For nuclear power stations, indirect GHG emissions are estimated by 8 to 120 g CO_2-eq./kWh$_{el}$. The number depends on concrete conditions. A global mean value is below 40 CO_2-eq./kWh$_{el}$ which is in the range of renewable sources (IPCC, 2007c). For German nuclear power stations a mean value of 32 g CO_2-eq./kWh is estimated compared to Russia and USA with 65 and 62 g CO_2-eq./kWh, respectively. This is more than the emissions of a highly efficient modern local power station for electricity production from natural gas using waste heat for district heating which is emitting 49 g CO_2-eq./kWh (Fritsche, 2007).

- For photovoltaic cells indirect GHG emissions of about 100 to 350 g CO_2-eq./kWh were established due to the very high production. Moreover real costs of CO_2 avoidance have to be considered. In case of photovoltaics costs are estimated at about EUR 1,000 per ton of CO_2 avoided as compared with EUR 40 to 45 for power stations and 20 to 30 for waste to energy plants.

- Geothermal energy is also not CO_2 neutral, since small quantities of CO_2 are normally released from the geological formations tapped for this energy form.

- For hydropower stations emissions from the water reservoir must be taken into account. In the reservoir, the organic matter is decomposed under anaerobic conditions, which results in methane production. The productivity depends on local climate conditions; it is much higher in warmer regions, but in general it is considered a serious problem. Several Brazilian hydro-reservoirs were studied using LCA (see chapter 8). For most projects LCA has shown low overall net GHG emissions (IPCC, 2007c). As an extremely negative example in Brazil, the GHG emissions from the Curuá-Una-Dam were measured 3.5 times that of electricity produced from oil (Graham-Rowe, 2005).

13.5 Climate Effects of Biomass Derived Fuels

Fossil fuels can be substituted by biogenic fuels, which may be derived from agriculture and forestry as biomass directly, or via the waste route from wasted material of biogenic origin. For principal routes see figure 13.4.

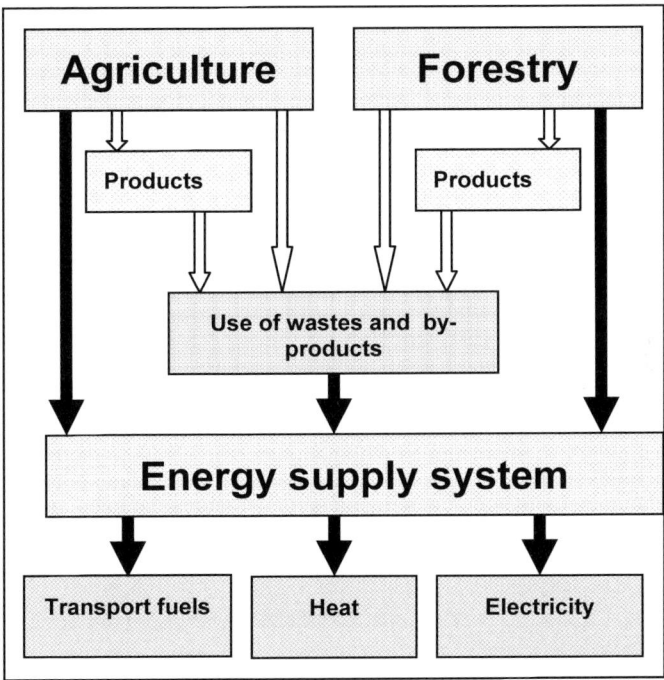

Figure 13.4 **Principal pathways from biogenic material into the energy system (after Öko-Institut, 2005)**

The combustion of biomass, biomass-derived fuels, or biogenic wastes which are used directly or as bioethanol, biodiesel, and biogas results in greenhouse gas emissions. But these are not addressed to climate change. As emissions are from biogenic material (and if the materials are grown on a sustainable basis), those emissions are considered to close the loop of the natural carbon cycle: They originate from atmospheric carbon dioxide from which they were taken by photosynthesis. CO_2 from biogenic sources will under natural conditions return to the atmosphere in the same quantity. Only if there are metabolites other than CO_2, such as methane, N_2O, NO_x, etc. then these substances are considered climate relevant and their greenhouse gas effects must be balanced.

Though positive effects of biofuels are of primary importance due to the reduction of greenhouse gas emissions, at the same time the reduction of the demands of fossil material sources is meaningful, and hence, the enhancement of the lifetime of this scarce and expensive resource. Moreover, biofuels support the development of agriculture in rural region, broaden the activity field of farmers, and opens new markets for products such as sugar. In many cases, residues from agriculture as well as from industry can be used as raw material sources of biofuels thus also contributing to reduce negative environmental effects of poor waste management.

However the intended positive effects will not fully occur in every case, and negative effects also have to be considered: Some biofuels are not grown environmental-friendly because they need pesticides and fertilizers which besides production efforts may cause water and air pollution. Impacts on biodiversity are of concern in the case of monocultures for renewable crops. In the international context, the main environmental risks are likely to be those concerning any large expansion in biofuel feedstock production, particularly in Brazil for sugar cane and in South East Asia for palm oil plantations. Growing demand for palm oil may be effectively contributing to clearance of rainforest in countries like Malaysia and Indonesia (Bauen, 2005). Moreover, energy is needed to grow the plants and to produce the biofuels from the crop. CO_2 is emitted out of the soil (see chapter 10.4); after the use of nitrogen fertilizers considerable emissions of N_2O occur.

Facing the world's rising food prices and recent shortages in nutritional crops, as another factor of concern the competition of energy and nutritional crops must be considered and balanced in such a way that primary needs are covered first. From a sustainability aspect these facts speak for a detailed complex analysis of all environmental, social and economical consequences of biofuels to prevent negative total effects.

13.5.1 Biofuels – facts and definitions

As figure 13.4 displays energy from biogenic resources may be applied as direct heating material, as a transport fuel or as a source of electricity in power stations. The following chapter deals with biofuels in the transport sector.

Types of biofuels are presented in table 13.5.

Table 13.5 Biofuel types and market products

Basic definitions	
Biofuel	Solid, liquid or gaseous fuel produced from plant or animal organic matter (biomass)
Biomass	Biodegradable fraction of products, waste and residues from agriculture including vegetal and animal substances, forestry and related industries, as well the biodegradable fraction of industrial and municipal solid waste
Synthetic biofuels	Synthetic hydrocarbons or mixtures of it produced from biomass, e.g. SynGas produced via gasification of forestry biomass, or SynDiesel
2nd generation biofuels	Biofuels such as bioethanol and biodiesel derived from lingo-cellulosic biomass by chemical or biological processes, especially by Fischer-Tropsch synthesis via gasification of biomass
Liquid biofuels	
Bioethanol	Ethanol produced from biomass and/or from the biodegradable fraction of waste, for use as biofuel. Most ethanol used for fuel is being blended into gasoline at concentrations of 5 to 10 percent. Fuel specification: E"x" contains x percent ethanol and (100-x) percent petrol, e.g. E5 and E85 with 5 and 85 percent of ethanol, respectively. E100 is non-blended bioethanol
Biodiesel	A methyl-ester produced from vegetable oil, animal oil, or recycled fats and oils of diesel quality, for use as biofuel. Specification: B"y" contains y percent biodiesel and (100-y) percent petroleum-based diesel, e.g. B5, B30, and B100 (non-blended biodiesel)
Biomethanol	Methanol produced from biomass, for use as biofuel
Bio-ETBE	Ethyl-Tertio-Butyl-Ether, produced from bioethanol, used as fuel additive to increase the octane rating and reduce knocking
Bio-MTBE	Methyl-Tertio-Butyl-Ether, produced from bio-methanol, used as fuel additive to increase the octane rating and reduce knocking
BtL	Biomass to Liquid (2nd generation biofuels)
Pure vegetable oil	Oil produced from oil plants through pressing, extraction or comparable procedures, chemically unmodified. Usable as biofuel if compatible with the type of engine involved and the corresponding emission requirements
Gaseous biofuels	
Bio-DME	Dimethylether produced from biomass for use as biofuel

Gaseous biofuels	
Biogas	A fuel gas produced from biomass and/or the biodegradable fraction of waste (in technical equipment, such as agricultural biogas plants which process manure; MBP technology; or as landfill gas), used as a fuel for power stations. Can be purified to natural gas quality
Bio-hydrogen	Hydrogen produced from biomass and/or the biodegradable fraction of waste for use as biofuel

Biofuels of the first generation are already being applied and will be more intensively introduced into the markets by 2010. Afterwards biofuels of the second generation will come into application from 2010 to 2020 as a result of intensive research which is already at the stage of demonstration projects (see table 13.6).

Table 13.6   Renewable fuels applicability timescale in the UK (Bauen, 2005)

Commercial applicability	Biofuel	Raw material sources
To 2010	Bioethanol	Starch and sugar crops: wheat grain, sugar beet, sugar cane, sorghum, corn
	Biodiesel	Oil crops and wastes: rapeseed, sunflower, soybean, palm oil, jatropha, waste vegetable oil, waste animal fat
	Biogas	Organic waste from agriculture (animal farming), wet energy crops
2010-2020	Biodiesel, bioethanol and biogas	Same as in the period up to 2010
	Bioethanol	Lignocellulose biomass: straw, wood, biodegradable municipal solid waste
	Fischer-Tropsch diesel	
	Hydrogen	Electrolysis of water using renewable electricity. Biomass feed-stocks (lignocellulose wastes, wet feed-stocks)

There is already considerable biofuel production capacity. It develops very progressively; growth rates of more than 10 percent per year are envisaged.

Bioethanol is the world's main biofuel which was produced in a quantity of around 36.5 Mio t in 2006 (BMELV, 2008). This represents two percent of global petrol use. Two countries, the USA and Brazil, produce more than three-quarters of world's total (see figure 13.5.).

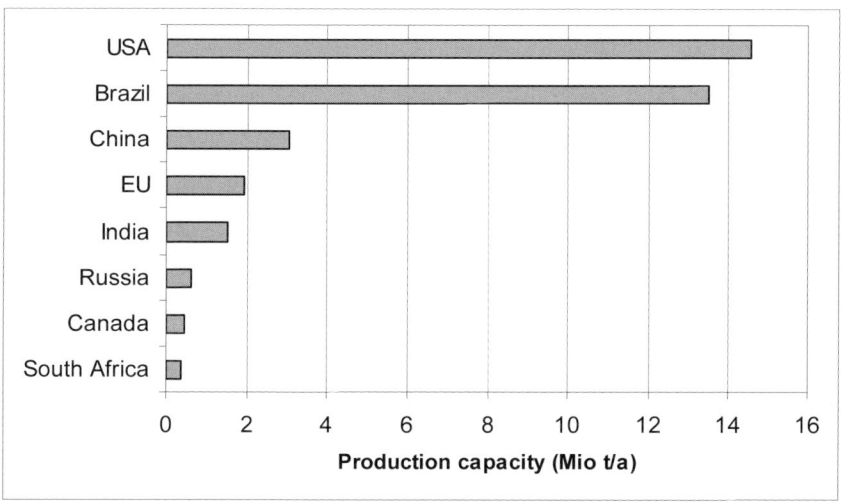

Figure 13.5   **Main ethanol production countries 2006 (BMELV, 2008)**

Biodiesel is by about 80 percent of the world's production from the EU (6.07 Mio t/a in 2006), with Germany the main producer with about 5.08 Mio t in 2007 (Bockey, 2007), equivant to 16 percent of the country's total diesel market, followed by Spain, France and Italy. The USA and Malaysia produced 0.83 and 0.60 Mio t/a, respectively (BMELV, 2008). Malaysia is the main producer of oils seeds for biodiesel production.

13.5.2   Current policies promoting biofuels

Most biofuels are still more costly than fossil fuels and thus production has to be encouraged e.g. by financial and organisational support. On the other hand measures are necessary to control negative environmental and social consequences such as shortage of nutritional crops, change in agricultural structures, or eradication of primeval forests.

Some examples of measures implemented in several countries are as follows:

- Most stimulating for biofuels was the Brazil *Proálcool* bioethanol programme, launched after the energy crisis of 1973. An extra aim was to use surplus sugar production. Sugar cane is the main source of bioethanol in Brazil. About two-thirds of the sugar cane biomass is processed into bioethanol. The commercial capacity is 400 facilities which currently produce a total of about 14 Mio t/a. This is now second largest national production in the world after decades in front-runner position. The costs of the ethanol production are in the range of 250 US$/m^3 which is less than half the costs of bioethanol in Europe (450 to 500 EUR/m^3). In Brazil currently all petrol is sold with an ethanol component of 20 to 26 percent (E25). One-half of the cars, a total of 14 Mio, are fuelled by E85 (see figure 13.6). Also pure ethanol (E100) is in use. Ethanol is 50 percent of the total national fuel market. A large share is exported. Refering to Biodiesel Brazil's current demand is 800 million liters. In 2008 the country imposed a two percent binding blend of biodiesel into regular diesel (B2). This is planned to go up to five percent (B5) in 2013.

Figure 13.6 **Ethanol fuelled test car in Brazil in the 1980th**

- The United States is now the world's largest bioethanol producer. Raw material is corn. A series of tax measures and incentives were followed by an exponential rise in production. In 2004 the *Volumetric Ethanol Excise Tax Credit (VEETC)* extended the existing tax incentives until 2010. In 2005, as part of a new energy bill, a *Renewable Fuels Standard (RFS)* was introduced, with a target rising production from 4 billion gallons in 2006 to 7.5 by 2012. EPA required a 3.71 percent blend of biofuels in 2007. The *Energy Independence and Security Act* which came into force in decembre 2007 includes a RFS calling for at least 36 billion gallons of ethanol to be used nationwide by 2022. Included is a national fuel economy standard of 35 miles per gallon by 2020, which will increase fuel economy standards by 40 percent. This will support the commercialization of bioethanol from cellulosic feedstocks such as native grasses, crop residues, forestry waste, and many other materials. The combination of domestically produced corn- and cellulose-based ethanol will replace a significant portion of fossil fuel based gasoline. Domestic oil use will be reduced by 5 million barrels per day by 2030. This will save fuel costs of about US$ 160 billion annually by 2030 and reduce GHG emissions by 320 Mio t CO_2-eq. annually (U.S., 2007).

- EU commission in its *Climate action and renewable energy package* issued on January 23, 2008, sets a mandatory EU target of 20 percent renewable energy by 2020 including a 10 percent biofuels target (EU, 2008). This target is subject to production being sustainable, 2nd generation biofuels becoming commercially available and the *Fuel Quality Directive* being amended accordingly to allow for adequate levels of blending. The Commission, after assessment of the impact of achieving this goal, concluded that it would need significant additional funds but result in a massive reduction of EU dependence on oil imports, generate employment and reduce GHG emissions by 35 percent. No renewable crops will be accepted in areas with a high carbon stock or a high biodiversity value.

- An enhanced contribution of renewables to the EU's total energy needs in the near future and the further development of the bioenergy market are very ambitious targets from an economic aspect, where investments into the technology of nearly EUR 165 billion are foreseen, and assuming that enough high yield cropland is available for energy crops instead of nutritional needs. Good agricultural practices including balancing of humus is necessary. In addition, future climate impacts on crops must be considered. Research provided evidence that the distribution of biofuel crops will increase in northern Europe and decrease in countries such as Spain, Portugal, southern France and Greece, due to increased drought. Mediterranean oil and solid biofuel crops will extend further north due to warmer summers. This shift will have dramatic consequences after 2080. The breeding of temperature and drought resistant plants is thus urgently necessary (Gill, 2006).

- In Germany biofuels in 2007 contributed 3.4 percent of total fuels, from which about 83 percent were biodiesel, and 8 and 10 percent were bioethanol and pure plant oil, respectively. A share of 16 percent of the total German diesel market originated from biodiesel. By *Biokraftstoffquotengesetz* (BMU, 2006) the share of biofuels in 2007 was fixed at 4.4 and 1.2 percent for diesel fuel and gasoline, respectively. In the case of gasoline, growth rate has to be 0.8 percent with a target biofuel amount of 3.6 percent in total by 2010. In addition, a total quota was defined for biofuels which will be 6.75 percent in 2010, and a linear growth of 8 percent by 2015. Regulation by EU Climate Action defines the 2020 target as 20 percent. All numbers are energy related, not mass related. In a *Roadmap Biokraftstoffe* (BMU, 2007) between the Automotive Industry and the German government it was agreed to promote and implement biofuels into transportation energy strategies, according to German and EU quota regulations. New fuel qualities with higher blends of biofuels will be defined, 2nd generation biofuels will be promoted by R&D, demonstration facilities, as well as implementation activities, and international proposals will be prepared for a sustainable plantation of renewable crops. However the increasing fuel tax will propably reduce marketing chances for this type of fuel.

- A number of non EU member countries have set targets for the incorporation of biofuels into conventional fuel. Amongst the measures, there are mandatory or fixed mixing percentages (e.g. Brazil, 25 percent mandatory; Canada: 3.5 percent target for 2010; in Ontario 5 percent for 2007), tax credits or incentives to biofuel producers or reduced fuel taxes (India: purchasing policy, 5 percent oil from indigenous plants in 2006, rising to 20 percent in 2020), or tax credits for vehicles (Brazil and Thailand: for cars run on biofuels; Thailand supports the development of a "Green vehicle") (EC, 2006).

13.5.3 Budgeting of climate consequences of biofuels

Ecological budgeting of bioethanol and biodiesel covers a wide range of GHG emissions according to the specific conditions on site. Figure 13.7 represents the results of a literature survey (Bauen, 2005). The difference between biofuels and fossil based petrol or diesel is displayed. Maximum and minimum values are given; GHG emissions of actual cases would lie between the two extremes.

In general the results elucidate that the GHG emissions range is between negative values and more than 100 percent improvement. Negative values indicate that more GHG emissions would occur compared to fossil fuel production. A value of more than 100 percent means that not only the GHG emissions from fossil fuels are compensated but a bonus results from benefits of by-products such as renewable energy generated during fuel production. In some cases there is more energy transformed into power than into resulting biofuel.

13.5.3.1 Climate impacts of bioethanol

For bioethanol, in figure 13.7 five variants of feed-stocks are shown with results ranging from minus 30 to 110 percent. The worst case is corn. Grain and sugar beet follow. Sugar cane and wood are best suited. The positive results using wood as a feedstock are for the gasification of wood and the transformation of the resulting process gas into ethanol by a Fischer-Tropsch synthesis (2nd generation biofuels – see table 13.5). On average about 60 percent as compared with fossil-fuel generated CO_2 emissions would result if bioethanol is applied.

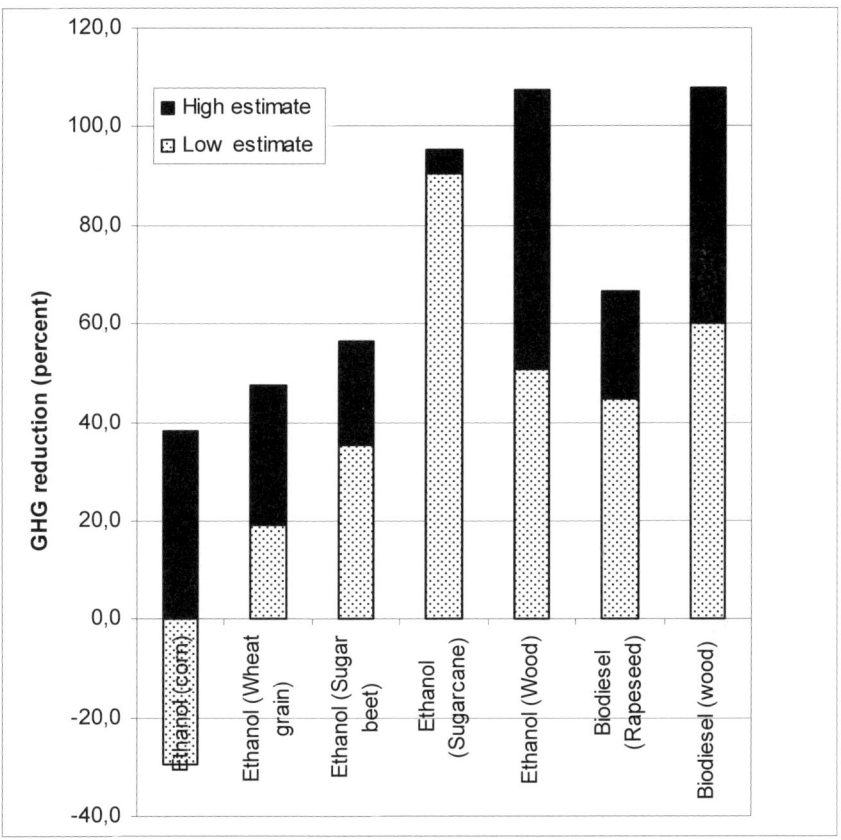

Figure 13.7   GHG effects of biofuel production (Bauen, 2005)

Bioethanol production from corn as a feed-stock is connected with high inputs of agrochemicals and pesticides, but also with high power needs for the production. Wheat and sugar beet are in the medium range.

Sugar cane is the basic feed-stock for the Brazilian *Proálcool* bioethanol programme. This is due to favourable cultivation conditions which pre-dominate in the Brazilian climate where a large part of solar radiation is transformed into plant biomass and hence into the energy source of bioethanol production. An energy output of bioethanol of about 85.5 GJ/ha from energy input of nearly 32.7 GJ per hectare (see figure 13.8) was reported in 1990. This is an energy input/energy output ratio of 2.6:1. Recently a much better ratio of 8.3 was documented for Brazil (Szwarc, 2004).

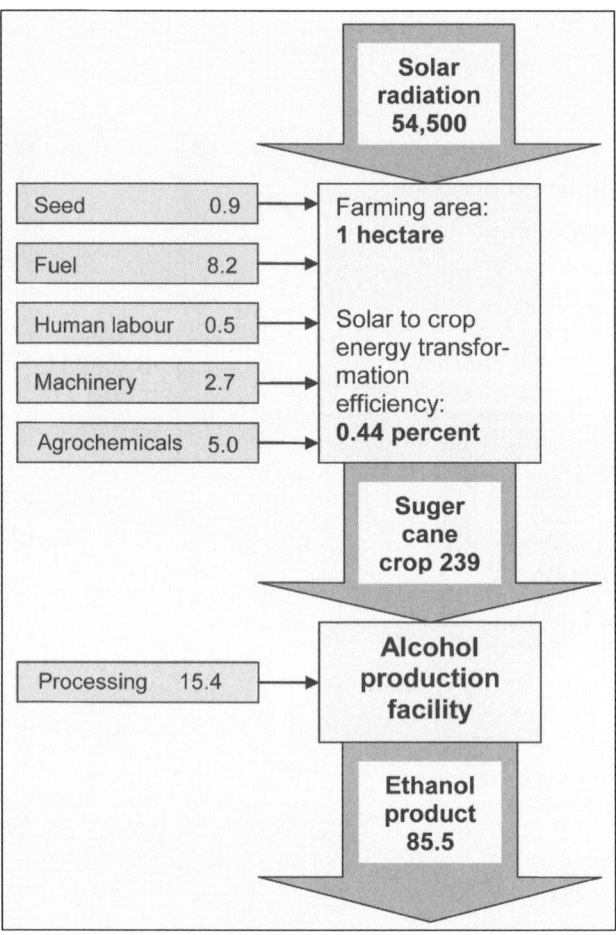

Figure 13.8   Energy budget of bioethanol in Brazil (in GJ/ha) (Soyez, 1990)

The LCA after a "well-to-wheel" approach indicates that even the GHG efforts for the long distance transport from South America to Europe do not compensate for the positive effects.

A more detailed analysis of the GHG emissions of the bioethanol production from wheat is given in table 13.7. It contains every process step and a range of

Climate Effects of Biomass Derived Fuels

concrete GHG emission values as well as the key variables which influence the process result.

Table 13.7 Emission sources in bioethanol production (Bauen, 2005)

Source of emission	Emission kg CO_2-eq./t bioethanol		Key variables
	minimal	maximal	
Feedstock-production			
Land use change	0	>1,000	Type of vegetation replaced (only significant where deforestation or vegetation changes occur)
Fertilizer manufacture	0	450	Type of fertilizer, fertilizing regime, crop yield, co-products
Emissions from soil	0	100	Soil conditions, climate, fertilizer applied, co-products
Fossil fuel used for cultivation	60	180	Tillage methods, tractor efficiency, co-products
Fossil fuel used for drying and storage	10	100	Farm equipment, energy used for drying, co-products
Transportation and processing			
Fossil fuel for transportation	20	50	Distance from farm, mode of transport 0, if renewable fuels are used
Transportation of product (Bioethanol)	20	80	Distance from farm to process, mode of transport
Processing			
Fossil fuel used for processing such as crushing, cleaning, drying	50	250	Type of crusher, moisture content, fuel used to power crusher, co-products
Hydrolysis, fermentation, distillation	-700	550	Type of process, export of heat and/or electricity
Total range	-540	>2,900	
Unleaded gasoline		3,135	

The resulting GHG emissions range from -540 to more than 2,900 kg CO_2-eq./t bioethanol produced. Extremes are unlikely but possible. Typical values for bioethanol from wheat under British conditions are given in figure 13.9. They sum up to about 1.250 kg CO_2-eq./t which is about 40 percent the value of gasoline.

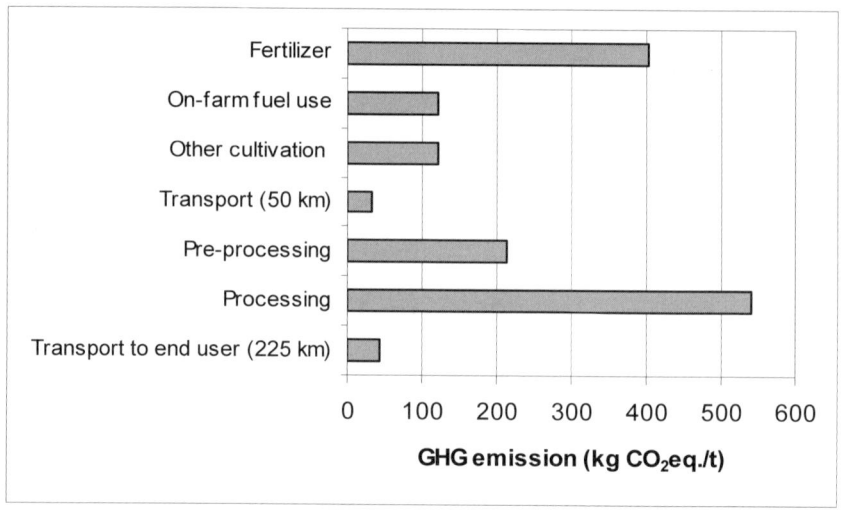

Figure 13.9   Typical GHG emissions for bioethanol from wheat (Bauen, 2005)

13.5.3.2   Climate impacts of biodiesel

With biodiesel the improvements relative to fossil fuel even in the worse case are in the range of more than 50 percent (see figure 13.7). The best result, with effects of more than 100 percent, would be reached if biodiesel was produced from wood as a feedstock, via the Fischer-Tropsch synthesis, where by-products of the process, such as electricity and heat, are considered.

For rapeseed as a feedstock of biodiesel (as RME) the life cycle assessment comes down with benefits for the greenhouse gas balance. Under German conditions, in the case of a 100,000 t plant, the benefit in GWP is 162 g CO_2-eq. per kWh compared to fossil diesel (BMELV, 2008).

Figure 13.10 displays the distance related effect of the use of biodiesel instead of fossil based diesel. In the case of the greenhouse gas potential the emissions of about 6 litres of diesel are avoided by application of RME at a distance of 100 km driven. In absolute numbers this amounts about 2.2 kg CO_2-eq./l RME (not shown in the figure). This means a clear climate related advantage for biodiesel.

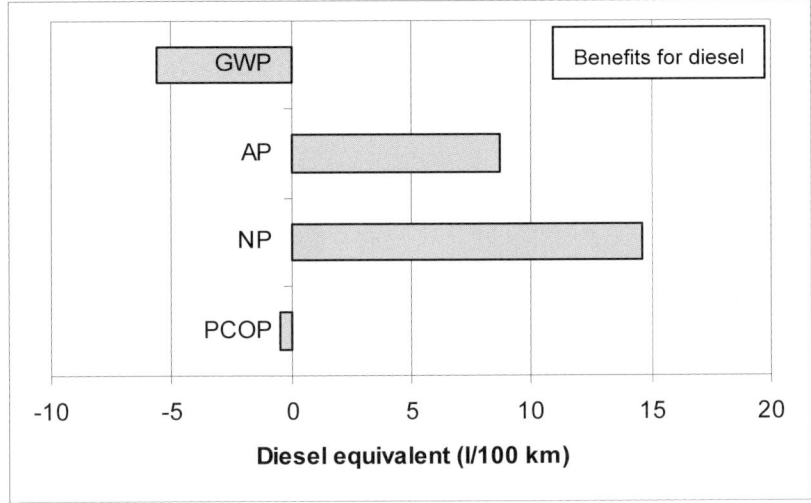

Figure 13.10   LCA of Biodiesel (RME) compared to fossil diesel (IFEU, 2003)

However nutrition and acidification potentials come up with disadvantages for biodiesel. With photochemical ozone formation no clear advantage can be seen. Therefore, considering all effects, a simple decision pro biodiesel is not possible. Only would biodiesel be the preferred solution if climate effects alone are considered important.

13.5.4   Consequences for biofuels application

A hundred percent reduction of greenhouse gas emissions will not be possible if biofuels of the first generation are applied. Better results would occur if, besides biofuel production, also power production during a combined process is envisaged. Thus instead of the simple replacement of fuels combined technological developments are to be supported.

If a replacement is envisioned in a first approach, it will be best to focus on such feed-stocks which result in a more than 50 percent emission reduction throughout the total supply chain. In future 80 percent reduction should be envisaged as the target. Moreover, the production of biofuels should also fulfill criteria concerning the origin, the production chain, and social aspects, especially when imported from less developed countries (EEB, 2005), as well as a full LCA.

Only with 2nd generation biofuels is a mostly greenhouse gas free fuel-supply possible which also does not need fossil fuel inputs (Picard, 2006).

14 Individual Activities to Reduce Climate Impacts

Information contained in the previous chapters may have given an insight into the climate change problems and of the approaches to reduce climate impacts in industry, energy production, waste management, or agriculture. It is the task of the managers in the companies, the scientists and the engineers, the waste managers, or the farmers to start activities to improve the situation in their field of responsibility.

However the impacts of climate on production and related processes represent only one part of the total problem. Another part is the involvement of individuals according to their life style and their behaviour as a consumer. The following chapter deals with two aspects: The one is to clarify to which extent production processes and individual consumption are related to a product's total climate impact during the life time from cradle to grave. The other is the affect on the individual behaviour of people in daily life.

14.1 Climate Impacts of Production and Consumption of Goods

An analysis of the total product life time "from cradle to grave" comes up with a surprising result: Consumers in many cases contribute to a much larger share to GHG emissions compared to production. This is elucidated by the following example which deals with GHG effects of household products (see figure 14.1).

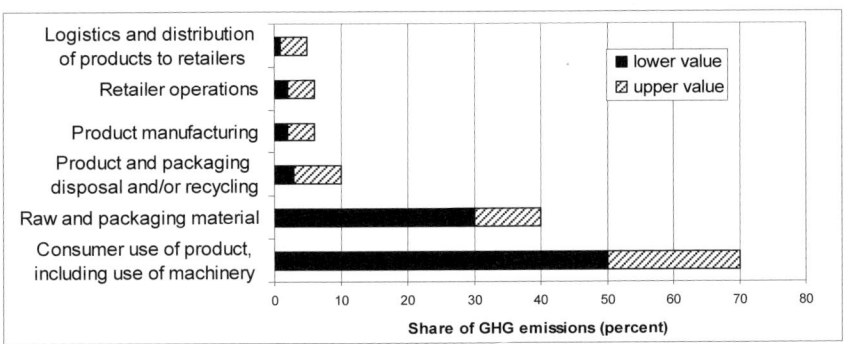

Figure 14.1 Influence of production and consumption of household products on GHG emissions (Reckitt, 2007)

Obviously the influence on climate of consumers is much larger than of manufacturing. In the case presented this is 50 to 70 percent compared to 3 to 7 percent. Even considering this as an extreme example it makes clear that behaviour of people is generally of great importance. A climate oriented individual life style is a key goal in tackling climate change.

14.2 Climate Impacts of Modern City Life Style

The life style in a city is characterized by activities such as public and private traffic; building and infrastructure; water, waste water and waste management; energy consumption; individual consumption of the people, and others. These activities are accompanied by climate impacts. The extent of these activities and thus the climate impact will depend on the economic, social, and ecological situation.

An example of typical everyday life greenhouse gas emissions in a small city in an industralized country is given in figure 14.2.

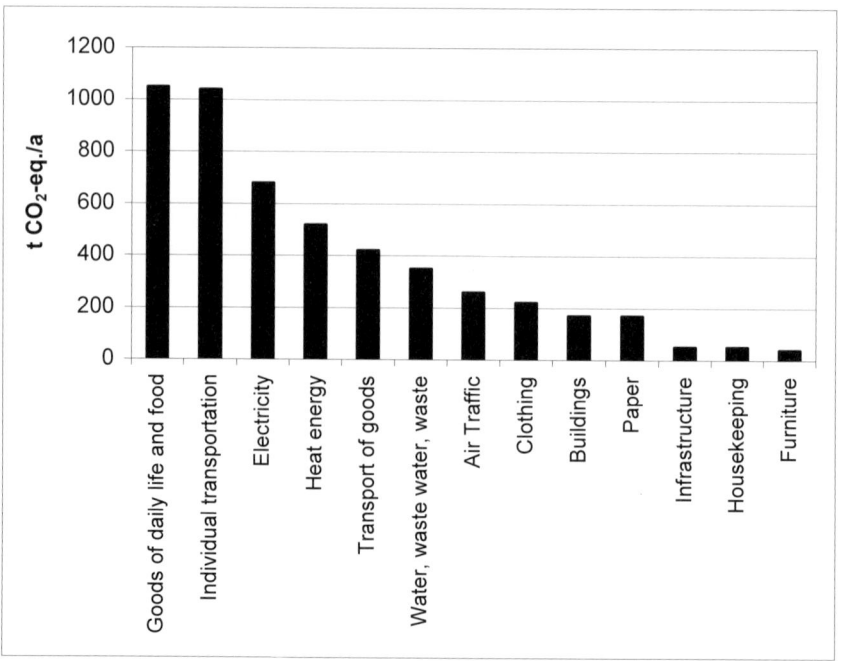

Figure 14.2   GHG emissions in a small German city (after Brohmann, 2002)

The largest contributors to climate impacts are the consumption of every day goods which is dominated by food and individual transportation. Electricity and heat consumption are at a relatively high level. Climate impacts of house building and infrastructure are marginal only.

In the case presented in figure 14.2 some measures were realized to reduce greenhouse gas emissions (see table 14.1).

Table 14.1   Results of a GHG reduction project in a German city (Brohmann, 2002)

Field of activity	Activity	Improvement (Percent)
Individual traffic	Implementation of an improved public traffic programme	22
Room heating	Changed energy standard; application of wood as fuel, and solar heating	92
Electricity	Application of photovoltaic electricity production	77
	Implementation of a power saving programme	1
Buildings	Use of wood as construction material	39
	Renovation of existing buildings instead of demolition and new construction	88

The best effects at a high starting level were achieved by the reduction of individual transportation through an improved public traffic concept in the city. Measures in room heating and electricity also resulted in large improvements, but from a much lower starting point. The large effects in buildings resulted in only small improvements due to the low starting level of GHG emissions in this sector.

14.3   Climate Oriented Individual Behaviour

Only recently policy makers have recognised the need to integrate thinking about climate change into all areas of public policy making, not only on the national and international levels, but on a regional or even individual level which meets the typical situation of everybody. The following two examples describe activities in the USA and the EU. In many other countries and regions similar proposals were made.

U.S. environmental protection agency EPA proposes a few simple things each of us can do for the reduction of climate burdens which focus on four main topics containing *Reduce, re-use, recycle*; *Reduce fossil fuel consumption*; *Tune up Your home* and *Get involved in public and at work* (EPA, 2007). Some examples are presented in table 14.2.

The *European Commission* in 2006 launched a campaign "*You Control Climate Change*" (EU, 2006b). The campaign aims to help interested individuals make a contribution to the fight against climate change. Small changes to the daily routine are envisaged by which significant reductions in greenhouse gas emissions and climate impacts will be possible if practised by many. This can be done by very simple everyday actions addressing the people under the slogan "*Turn down, Switch off, Recycle, Walk*". Table 14.2 displays a set of personal climate control actions and quantifies possible impact reduction. A calculation programme named "*carbon-calculator*" provides detailed background information and automatically checks the effects of personal activities (EC, 2006a).

Table 14.2   Personal climate control activities (EC, 2006)

Activity	kg CO_2-eq./a
Turn down	
Insulate your home	630
Programme thermostat temperature low at night or when out of the house	440
Replace your old single-glazed windows with double-glazed ones	350
Reduce the temperature of your home by just 1°C	300
When replacing your old, buy a fridgefreezer with European Grade A+ (with automatic defrost cycles)	210
Place your fridge in a cool environment	150
Let hot food cool down before putting it in the fridge	6
Switch off	
Switch to green electricity	520
Go solar. Install a solar thermal system in your home to help provide your hot water (EPA, 2007)	400
Switch off your air conditioner when you're not in the house in summertime (Example: 5 hours switched off)	300

Activity	kg CO_2-eq./a
Air dry clothes naturally instead of using a tumble dryer	280
Switch off lights in hallways and rooms of your house when you don't need them (Example: Five 60W lights)	270
Replace ordinary light bulbs in lamps with low energy ones (Example: 5 lamps 5 hours a day)	250
Install a low-flow showerhead	230
Use the washing machine only when it's full	45
Plug voltage transformers into a socket with a switch (Example: 4 transformers)	35
When replacing your old television buy an eco-labelled one	30
Boil just enough water for your hot drink	25
Make sure to turn off or fix dripping taps	20
Turn off the tap while brushing your teeth	3
Recycle	
Use recycling systems provided by your waste management company. (Example: Recycling all of your home's waste newsprint, cardboard, glass, metal, and organics) (EPA, 2007)	500
Buying bottled beverages: buy big bottles instead of the equivalent amount of small ones (Example: one 1,5 l bottle instead of three 0,5 l)	9
Use a reusable bag each time you go shopping	8
Consider whether a document or e-mail must really be printed	7
Recycle your waste (Example: recycle 1 kg of aluminium, see table 12.2)	6
Walk	
Think about giving your car a day off. Consider transportation alternatives such as mass transit, carpooling, bicycling, and telecommunicating (Example: Leaving your car at home two days a week) (EPA, 2007)	800
Walk	
When replacing your old car with a new one, consider the fuel need	660
Skip flights (Example: Per year 2 flights of 1.5 hours distance)	270

Activity	kg CO$_2$-eq./a
Walk	
Replace your short car journeys with biking	240
Apply smart driving: Plan your journey, start your car moving without pressing down the throttle, shift to a higher gear as soon as possible (at 2000-2500 rpm), keep the speed steady, and look ahead to avoid sudden breaking and accelerating. Turn off the engine even at short stops!	200
Make sure you have correct tyre pressure in your car	140
Travel by train instead of travelling by car on your own (Example: 1000 km per year)	110
Change to low viscosity motor oil	45
Reduce speed (Example: from 110 to 90 km/h for 10 percent of driving distance)	35

Literature

Albrecht, B.A. (1989): Aerosols, cloud microphysics, and fractional cloudiness. Science, 245, 1227-1230.

Allianz (2003): Umweltschonend mobil. Bahn, Auto, Flugzeug, Schiff im Umweltvergleich. Allianz pro Schiene e.V., 2003. http://www.allianz-pro-schiene.de/sites/ApS_UmweltschonendMobil_001203.pdf

Amelung, D. (2007): Stahlmarkt in Bestform. www.stahlonline.de/medien_lounge/vortraege/1604PKHannoverMesse.2007Text.pdf

BASF (2004): Eco-efficiency analysis. http://corporate.basf.com/en/sustainability/oekoeffizienz/?id=noAIH8fMmbcp1tZ

Bauen, A., et al. (2005): Feasibility study on certification for a renewable transport fuel obligation. http://www.dft.gov.uk/stellent/groups/dft_roads/documents/page/dft_roads_610329-03.hcsp#P37_6185

Bengtsson, L., I. Hodges and E. Roeckner (2006): Storm tracks and climate change. Journal of Climate, 19, 3518-3543.

BMELV (Bundesministerium für Ernährung, Landwirtschaft und Verbraucherschutz) (2008): Nutzung von Biomasse für die Energiegewinnung. Wissenschaftlicher Beirat Agrarpolitik, Berlin, November 2007. http://www.bmelv.de

BMU (Bundesministerium für Umwelt, Naturschutz und Reaktorsicherheit) (2001): Ordinance on environmentally compatible storage of waste from human settlements and on biological waste-treatment facilities, Berlin 2001.

BMU (Bundesministerium für Umwelt, Naturschutz und Reaktorsicherheit) (2007): Roadmap Biokraftstoffe. http://www.bmu.de/files/pdfs/allgemein/application/pdf/roadmap_biokraftstoffe.pdf

BMWi (Federal Ministry of Economics and Technology) (2007): CCS-Technologien in Deutschland. Entwicklungsstand und Perspektiven. http://www.bmwi.de/BMWi/Redaktion/PDF/B/bericht-entwicklungsstand-und-perspektiven-von-ccs-technologien-in-deutschland,property=pdf,bereich=bmwi,sprache=de,rwb=true.pdf

Bockey, D. (2007): Biodiesel & Co. UFOP-Bericht 2006/07. http://www.ufop.de/downloads/Biodiesel_Co_081007.pdf

BR&D (Bio-climate Research and Development) (2005): Biofuels Carbon Declaration Methodology.
http://www.greenergy.com/expertise/pdfs/Carbon_Declaration_Methodology.pdf

BREF (2002): Reference document for the best available technique reference document Iron and Steel. European Commission.
http://www.wbcsdcement.org/pdf/tf1/tf1_guidelines.pdf

BREF (2006): Reference document on the best available techniques for waste incineration. European Commission.
http://www.bvt.umweltbundesamt.de/archiv-e/waste_incineration.pdf

Bringezu, S. (2000): Ressourcennutzung in Wirtschaftsräumen. Stoffstromanalysen für eine nachhaltige Raumentwicklung. Berlin, 2000.

Brohmann, B., Fritsche, U., et al. (2002): Nachhaltige Stadtteile auf innerstädtischen Konversionsflächen: Stoffstromanalyse als Bewertungsinstrument. IFEU-Institut. Darmstadt, 2002. http://www.oeko.de/service/cities.

Brunner, P.H., Rechberger, H. (2004): Practical handbook on material flow analysis. Boca Raton, Florida, 2004.

CDP (Carbon Disclosure Project) (2007): Carbon disclosure project.
http://www.cdproject.net

CEN (European Committee for Standardization) (1999): (CEN)-ISO 14040-14043. Brussels, 1997-1999.

Crutzen, P.J. (2006): Albedo enhancement by stratospheric sulfur injections: A contribution to resolve a policy dilemma? Climatic Change, doi 10.1007/s10584-006-9101-y.

CSI (2005): Cement sustainability initiative: The Cement CO_2 Protocol.
http://www.wbcsdcement.org/pdf/tf1/tf1_guidelines.pdf

CSIRO (2001): Emissions: Graziers Flock To Block Burps. *ScienceDaily*. Retrieved February 13, 2008, from
http://www.sciencedaily.com/releases/ 2001/06/010611071759.htm

Devasthale, A. (2005): Aerosol indirect effect in the thermal spectral range as seen from satellites. Ph.D. thesis, University of Hamburg, Hamburg Reports on Earth System Science/Bericht zur Erdsystemforschung 16.

Devasthale, A., et al. (2005b): Change in cloud-top temperatures over Europe. IEEE Geoscience and Remote Sensing Letters, 2, 333-336.

Devasthale, A., et al. (2006): Impact of ship emissions on cloud properties over coastal areas. Geophys. Res. Letters, 33, Seq. No. L02811, doi:10.1029/2005GL024470.

Doedens, H., Gallenkemper, B., et al. (2007): MBP – state of the art in Germany. Müll und Abfall (www.muellundabfall.de), 39(12)2007, p. 576-579

DSD (Duales System Deutschland) (2002): The Green Dot and its Benefit for the Environment. Study by Öko-Institut on commission for Duales System Deutschland AG. Cologne, March 2002.

DSD (Duales System Deutschland) (2002b): The green dot and its benefits for the environment. Cologne, 2002.

DSD (Duales System Deutschland) (2007): Significant increase in plastics recovery. Cologne, 2 May 2007. www.gruener-punkt.de

EADS (European Aeronautic Defence and Space Company) (2008): Airbus completes first test flight with alternative fuel on civil aircraft, Toulouse, 01 February 2008. http://www.eads.com/1024/en/pressdb/pressdb/Airbus/20080201_airbus_alternative_fuel_success.html

EC (European Commission) (2004): Working document on sludge and biowaste. Draft discussion document. Brussels, 15-16 January, 2004.

EC (European Commission) (2006): An EU strategy for biofuels. KOM (2006)34 final. Brussels, 8.2.2006 http://ec.europa.eu/energy/res/biomass_action_plan/doc/2006_02_08_comm_eu_strategy_en.pdf

EC (European Commission) (2006a): Carbon calculator. http://www.mycarbonfootprint.eu/carboncalculator1_en.asp

EC (European Commission) (2006b): Climate Campaign 2006. http://ec.europa.eu/environment/climat/campaign/index_en.htm

EC (European Commission) (2007): Annual European Community greenhouse gas inventory 1990-2005 and inventory report 2006. http://reports.eea.europa.eu/ technical_report_2006_6/en/EC-GHG-Inventory-2006.pdf, http://reports.eea.europa.eu/technical_report_2007_7/en/Full%20report%20Annual%20European%20Community%20greenhouse%20gas%20inventory%201990-2005%20and%20inventory%20report%202007.pdf

EC (European Commission) (2008): Climate action and renewable energy package, 23.01.2008: http://ec.europa.eu/environment/climat/climate_action.htm

EEA (European Environment Agency) (2008): Better management of municipal waste will reduce GHG emissions. EEA briefing 01, 2008. http://reports.eea.europa.eu/briefing_2008_1/en/EN_Briefing_01-2008.pdf

EEB (European Environmental Bureau) (2005): EEB position on Biomass and biofuels: the need of well defined sustainability criteria, December, 2005:

http://www.eeb.org/activities/agriculture/EEB-position-on-bioenergy-191205.pdf

EPA (2005): International Non-CO_2 greenhouse Gas marginal abatement report. Draft methane and nitrous from non-agricultural sources, April 2005.
http://www.epa.gov/methane/pdfs/chapter3_landfill.pdf

EPA (2005a): Calculating Greenhouse Gas Emissions with the Excel version of the WAste Reduction Model (WARM).
http://www.epa.gov/globalwarming/actions/waste/warm.htm and
http://yosemite.epa.gov/oar/globalwarming.nsf/content/ActionsWasteWARM.html

EPA (2005b): International non-CO_2 greenhouse gas marginal abatement report. Draft methane and nitrous from non-agricultural sources. Chapter 6. April, 2005.
http://www.epa.gov/methane/pdfs/chatper6_industrial_process.pdf

EPA (2006a): Global anthropogenic Non-CO_2 greenhouse gas emissions: 1990-2020. Revised. June, 2006.
http://www.epa.gov/climatechange/economics/downloads/GlobalAnthroEmissionsReport.pdf

EPA (2006b): U.S. emission inventory 2006.
http://yosemite.epa.gov/oar/globalwarming.nsf/content/ResourceCenterPublicationsGHGEmissionsUSEmissionsInventory2006.html

EPA (2006c): What You can do. Source reduction and recycling.
http://epa.gov/climatechange/wycd/waste/downloads/chapter3.pdf

EPA (2007): What can we do about global warming.
http://yosemite.epa.gov/oar/globalwarming.nsf/UniqueKeyLookup/SHSU5BWJKS/$File/whatcanwedoaboutgw.pdf

EPA (2007b): Inventory of U.S. greenhouse gas emissions and sinks: 1990-2005. Washington, 2007.
http://www.epa.gov/climatechange/emissions/downloads06/07Trends.pdf

EPA (2007c): Climate Change – Health and Environmental Effects. Agriculture and Food Supply.
http://www.epa.gov/climatechange/effects/agriculture.html#ref

EPA (2007d): Landfill gas energy projects and candidate landfills.
http://www.epa.gov/lmop/docs/map.pdf

EPA (2007e): Methane to Markets. http://www.methanetomarkets.org

EPA (2007f): CMM global overview.
http://www.methanetomarkets.org/resources/coalmines/overview.htm

EPA (2007g): Methane to markets. Landfills. Basic information.
http://www.methanetomarkets.org/landfills/landfills-bkgrd.htm

EPA (2008): Press release, 15.01.2008: EPA Recognizes Initiatives for Harnessing Renewable Energy from Landfills.
http://yosemite.epa.gov/opa/admpress.nsf/d0cf6618525a9efb85257359003fb69d/9323a77157f6c927852573d10069b2a0! Open Document.

EPA (2008a): Inventory of U.S. greenhouse gas emissions and sinks: 1990-2006. Washington, 2008.
http://www.epa.gov/climatechange/emissions/downloads/08_CR.pdf

EPA (2008b): Why recycling some materials reduces GHG emissions more than source reduction.
http://www.epa.gov/climatechange/wycd/waste/calculators/SRvsRecycling.html

EU (European Union) (2005): Eco-design of energy-using products.
http://europa.eu.int/comm/enterprise/eco_design

Eunomia (2002): Economic analysis of options for managing biodegradable municipal waste. Final Report, Brussels.
http://ec.europa.eu/environment/waste/compost/pdf/econanalysis_finalreport.pdf

European Space Agency (ESA) (2006): The Changing Earth. ESA-SP-1304, ISBN 92-9092-457-8.

FAO (UN Food and Agricultural Organization) (2005): Major food and agricultural commodities and producers, 2005.
http://www.fao.org/es/ess/top/commodity.html?lang=en&item=27&year=2005

Fritsche, U., et al. (2007): Comparison of Greenhouse-Gas Emissions and Abatement Cost of Nuclear and Alternative Energy Options from a Life-Cycle Perspective. Öko-Institut, Darmstadt, 2007.
http://www.oeko.de/service/gemis/files/doku/2007akw_co2papier.pdf

FVB (Fachverband Biogas) (2006): Biogas-Fakten.
http://www.biogas.org/datenbank/file/notmember/medien/Fakten_Biogas_2006_03.pdf

Gill, T., et al. (2006): The potential distribution of bioenergy crops in Europe under the present and future climate. Biomass and bioenergy 30 (2006), 183-197.

Gohlke, O., Neukirchen, B., Wiesner, J. (Hrsg.) (2006): Werkzeuge zur Bewertung von Abfallbehandlungsverfahren – Methoden und Ergebnisse. VDI-Verlag, Düsseldorf, 2006.

Graham-Rowe, D. (2005): Hydroelectric power's dirty secret revealed. New Scientist, February, 26th, 2005, p. 8

Grassl, H. (1975): Albedo reduction and radiative heating of clouds by absorbing aerosol particles. Beitr. Phys. Atm., 48, 199-210.

Hagengut, C. (2002): Packaging Recovery and Recycling in Europe. Implementation of the European Directive on Packaging and Packaging Waste. Inventory 2001. INTEC consulting, Bonn, 2002.

Hasselmann, K. (1993): Optimal fingerprints for the detection of time-dependent climate change. J. of Climate, 6, 1957-1971

Hegerl, G.C., H. von Storch, K. Hasselmann, B.D. Santer, U. Cubasch, and P.D. Jones (1996): Detecting greenhouse-gas-induced climate change with an optimal fingerprint method. J. of Climate, 9, 2281-2306. http://www.cambridge.org/9780521700801.

IFEU (Institut für Energie- und Umweltforschung) (2003): Gutachten zur Erweiterung der Ökobilanz für RME. Heidelberg, 2003.

IPCC (2000): Special Report on Emission Scenarios. WMO, Geneva, Switzerland.

IPCC (2001a): Climate Change: The Scientific Basis. Contribution of Working Group I to the Third Assessment Report (TAR). Cambridge University Press, Cambridge, UK, 881 p.

IPCC (2001b): Impacts, Adaptation, and Vulnerability. Contribution of Working Group II to the Third Assessment Report. Cambridge University Press, Cambridge, UK.

IPCC (2005): IPCC special report "Carbon capture and storage". Montreal, 2005. http://arch.rivm.nl/env/int/ipcc/pages_media/SRCCS-final/SRCCS_SummaryforPolicymakers.pdf

IPCC (2007a): The Science of Climate Change, Working Group I Report, Fourth Assessment Report of IPCC, Cambridge University Press, Cambridge, UK.

IPCC (2007b): Fourth Assessment Report of IPCC, Working Group II Report, Cambridge University Press, Cambridge, UK.

IPCC (2007c): Bernstein, L., J. Roy, K. C. Delhotal, J. Harnisch, R. Matsuhashi, L. Price, K. Tanaka, E. Worrell, F. Yamba, Z. Fengqi, 2007: Industry. In Climate Change 2007: Mitigation. Contribution of Working Group III to the Fourth Assessment Report of the Intergovernmental Panel on Climate Change [B. Metz, O.R. Davidson, P.R. Bosch, R. Dave, L.A. Meyer (eds)], Cambridge University Press, Cambridge, United Kingdom and New York, NY, USA.
http://www.mnp.nl/ipcc/pages_media/FAR4docs/final_pdfs_ar4/Chapter07.pdf

Kageson, P. (2005): Reducing CO_2 emissions from new cars. http://www.transportenvironment.org/docs/Publications/2005pubs/05-1_te_co2_cars.pdf

Kley, D., J.M Russel III, and C. Phillips (2000): SPARC Assessment of Upper Tropospheric and Stratospheric Water Vapour. SPARC Report, No. 2, WMO, Geneva, Switzerland.

Knappe, F. (2004): Die Bioabfallverwertung aus ökologischer Sicht. In: Johnke, B., Scheffran, J., Soyez, K. (Eds): Abfall, Energie und Klima. Berlin.

Krueger, O. and H. Grassl (2004): Albedo reduction by absorbing aerosols over China. Geophysical Research Letters, Vol. 31, L02 018, L2108-L2112.

Krueger, O., H. Grassl (2002): The indirect aerosol effect over Europe. Geophysical Research Letters, Vol. 29, No. 19, 1925, doi:10.1029/2001GL014081, 2002.

Lehmann, J., J. Gaunt and M. Rondon (2006): Bio-char sequestration in terrestrial ecosystems – a review; Mitigation and Adaptation Strategies for Global Change, 11, 403-427.

Mandra, O., Novakivska, N. (2004): Leak reduction at natural gas compressor stations of the gas transition system of Ukraine. www.epa.gov/gasstar/workshops/houston-oct2004/mandra508.pdf

Marheineke,T., Friedrich, R. et al. (1999): Ganzheitliche Bilanzierung der Energie- und Stoffströme in Energieversorgungstechniken. Stuttgart, 1999. FZB-Berichte, Universität Stuttgart.

Markert, H. (2008): Technical description of the Naunhof biogas plant. www.biogas-markert.de

Marland, G., Boden, T.A., Andres, R.J. (2004): Global, regional, and national fossil fuel CO_2 emissions. http://cdiac.esd.ornl.gov/trends/emis/meth_reg.htm

MPI (Max Planck Institute for Meteorology) (2006a): Klimaprojektionen des 21. Jahrhunderts. 26 pages.

MPI (Max Planck Institute for Meteorology) (2006b): http://www.mpimet.mpg.de/wissenschaft/Überblick/Atmosphäre-im-Erdsystem/regionale-Klimamodellierunge.html.

Nelles, M., Degener, P. (2007): MBA technology from Germany. Müll und Abfall (www.muellundabfall.de), 39(2007)12, 605-610

Noack, H.-C. (2007): Jenseits von 350 km ist das Flugzeug ökologisch am besten. FAZ, Nr. 40, 16.2.07.

Öko-Institut (2005): Bioenergy. New growth for Germany. Heidelberg, 2005. http://www.oeko.de/service/bio/de/index.htm

Öko-Institut (2006): Global Emission Model for Integrated Systems (GEMIS), Heidelberg. http://www.oeko.de/service/gemis/en/index.htm

Picard, K. (2006): Biokraftstoffe aus Sicht der Mineralölindustrie. Technikfolgenabschätzung, 15(2006)1, 34-42.

Raes, F., T. Bates, F. Mc Govern, M. Van Liedekerke (2000): The second Aeorosol Characterization Experiment (ACE-2): General overview and main results. Tellus, 52B, 2, 111-125.

Ramanathan, V., et al. (2005): Atmospheric brown clouds: Impacts on South Asian climate and hydrological cycle. PNAS, 102, 5326-5333.

Reckitt (2007): Reckitt Benckiser targets new standard in carbon reduction. Carbon 20. Press release by Reckitt-Benckiser, Nov. 1, 2007. http://www.reckittbenckiser.com/Sites/Carbon20/index.html

RWE (2006): RWE plant weltweit erstes CO_2-freies Großkraftwerk für Kohle inclusive CO_2-Speicherung. http://www.rwe.com/generator.aspx/templateId=renderPage/id=76858?pmid=4000956

Sandholzer, D., Niederl, A., Narodoslawsky, M. (2005): SPIonExcel – fast and easy calculation of the Sustainable Process Index via computer. Chemical Engineering Transactions (2005), 7 (2), 443-446. http://www.spionexcel.tugraz.at

Schellnhuber, H.-J. (2001): Die Koevolution von Natur, Gesellschaft und Wissenschaft – Eine Dreiecksbeziehung wird kritisch. GAIA, 10, 258-262.

Schön, M. (2004): BINE-Informationsdienst II/04, energieintensive Grundstoffe – Effizienzpotenziale und Perspektiven. Karlsruhe.

Schön, M., et al. (1993): Emissionen der Treibhausgase Distickstoffoxid und Methan in Deutschland. Fraunhofer-Institut für Systemtechnik und Innovationsforschung, Forschungsbericht 93-121, Karlsruhe.

Schönau (2007): Elektrizitätswerke Schönau GmbH. Strom-Herkunftsnachweis 2006. http://www.ews-schoenau.de/Download/files/Herkunftsnachweis2006.pdf

Simeprodeso (2007): Project case study: Metropolitan solid waste processing system Landfill gas energy project Monterrey, Mexico. http://www.methanetomarkets.org/landfills/landfills-bkgrd.htm

Smith, A., Brown, K., et al. (2001): Waste management options and climate change. EC, 2001, http://europa.eu.int/comm/environment/waste/studies/climate_change.htm

Soyez, K. (1990): Biotechnologie. ISBN 3-7643-2248-9. Basel.

Soyez, K. (1990b): Kombiniertes Verfahren zur Schnellkompostierung, regenerativen CO_2-Düngung und Abwärmenutzung im Gartenbau. In: Dott, W. u.a. (Hrsg.): Biologische Verfahren der Abfallbehandlung. Berlin, 1990.

Soyez, K. (Hrsg.) (2001): Mechanisch-biologische Abfallbehandlung. Technologien, Ablagerungsverhalten, Bewertung. ISBN 3-503-060006. Berlin.

Soyez, K., Markert, H. (2008): Implementation potential of Biogas-Plants in South Africa. Müll und Abfall (to be published, 2008)

Soyez, K., Plickert, S. (2002): Mechanical-biological waste treatment. In: Ludwig, C., et al. (edts): Municipal solid waste management. ISBN 3-540-44100-X. Berlin.

Span, R. (2006): Thermophysical properties for analysis and design of power cycles with capture of carbon dioxide. http://thet.uni-paderborn.de/chronik/Boulder2006/Properties%20for%20CO2%20Capture.pdf

Stern, N. (2006): The Economics of Climate Change. ISBN-13: 9780521700801. http://www.cambridge.org/catalogue/catalogue.asp?isbn=9780521700801

Szwarc, A. (2004): Use of biofuels in Brazil. http://unfccc.int/files/meetings/cop_10/in_session_workshops/mitigation/application/pdf/041209szwarc-use_biofuels_in_brazil.pdf

T&E (European Federation for transport and environment) (2006): EU climate policy for passenger cars. Background briefing, August 2006. http://www.transportenvironment.org/docs/Publications/2006/2006-08_cars_co2_background_briefing.pdf

Twomey, S. (1974): Pollution and planetary albedo. Atm. Environment, 8, 1251-1256.

U.S. (United States Administration) (2007): Energy Independence and Security Act of 2007. http://www.whitehouse.gov/news/releases/2007/12/20071219-1.html

UBA (Federal Environment Agency) (2007): National inventory report for the German greenhouse gas inventory 1990-2007. Dessau. www.umweltbundesamt.de/emissionen/archiv/nir_2007_e.pdf

UNEP (United Nations Environment Program) (2007): International life cycle partnership. www.unep.fr/pc/sustain/lcinitiative/home.htm

UNFCCC (United Nations Framework Convention on Climate Change) (2007): Inventory reports of annex 1 states. http://unfccc.int/national_reports/annex_i_ghg_inventories/national_inventories_submissions/items/3929.php

USGS (U.S. Geological Survey) (2007): Commodity statistics and information. http://minerals.usgs.gov/minerals/pubs/commodity

Van de Sand, J., Bals, C. (2005): Germanwatch: Hintergrundpapier. Deutsche Autoindustrie klagt gegen Klimaschutzgesetz Kaliforniens. http://www.germanwatch.org/rio/auto05hg.pdf

Vattenfall (2007): Climate map 2030. www.vattenfall.com/climatemap

VHK (Van Holsteijn en Kemma BV) (2005): Methodology Study Eco-Design of energy using Products. Final Report, Delft, November 2005. www.eupproject.org

Vizcaino, M. (2006): Long-term interactions between ice sheets and climate under anthropogenic greenhouse forcing – Simulations with two complex Earth system models. Ph.D thesis, University of Hamburg, Hamburg.

VWEW (Verband der Elektrizitätswerke e.V.) (2006): Strommarkt Deutschland. Zahlen und Fakten zur Stromversorgung.

Wassmann R., et al. (eds) (2000): Methane Emissions from Major Rice Ecosystems in Asia. Nutr. Cycling Agroecosyst. 58. 2000.

WBGU (Wissenschaftlicher Beirat der Bundesregierung „Globale Umweltveränderungen) (2003a): Welt im Wandel – Energiewende zur Nachhaltigkeit, Springer Verlag Berlin – Heidelberg, ISBN 2-540-40160-1.

WBGU (Wissenschaftlicher Beirat der Bundesregierung „Globale Umweltveränderungen) (2003b): Über Kioto hinaus denken – Klimaschutzstrategien für das 21. Jahrhundert, WBGU, Berlin, ISBN 3-936191-03-4.

WBGU (Wissenschaftlicher Beirat der Bundesregierung „Globale Umweltveränderungen) (2003a): Welt im Wandel – Energiewende zur Nachhaltigkeit, Springer Verlag Berlin – Heidelberg, ISBN 2-540-40160-1.

WBGU (Wissenschaftlicher Beirat der Bundesregierung „Globale Umweltveränderungen) (2003b): Über Kioto hinaus denken – Klimaschutzstrategien für das 21. Jahrhundert, WBGU, Berlin, ISBN 3-936191-03-4.

WBGU (Wissenschaftlicher Beirat der Bundesregierung „Globale Umweltveränderungen" (2006): Die Zukunft der Meeres – zu warm, zu hoch, zu sauer. ISBN 3-936191-13-1.

WWF (formerly known as World Wildlife Fund) (2006): Power switch campaign: http://powerswitch.panda.org/responsible/coal_power_dirty.cfm

Zeschmar-Lahl, B. (2000): Thermisch-regenerative Abgasreinigung für die mechanisch-biologische Abfallbehandlung. BZL, Oyten, 2000.

Indication of Sources in Subtitles of Figures

Fig. 2.1: Hartmut Grassl (2000)

Fig. 2.3: IPCC (2007a), bearbeitet von MPI-M (2007)

Fig. 2.4: IPCC, Summary for Policymakers (SPM) (2007a)

Fig. 2.8: Lennart Bengtsson (2005)

Fig. 3.2: Filippo Giorgi (2006)

Fig. 3.5: MPI-M (2006a)

Fig. 3.7: MPI-M (2006a)

Fig. 3.9: MPI-M (2006b)

Fig. 3.11: MPI-M (2006b)

Fig. 5.1: WBGU (2003b)

Fig. 6.3: WBGU (2003a)

Fig. 6.4: WBGU (2003a)

Fig. 6.5: WBGU (2003b)

Fig. 11.3: Ingenieurbüro Dr. Markert, Kaltennordheim/Rhön, Germany

Fig. 13.6: Volkswagen AG, Wolfsburg, Germany